SOTHIS PRESS

Sothis Press aims to completely reimagine the public domain book with an emphasis on universal access to captivatingly designed print editions. We believe that technology will enable every single book ever published to be perpetually accessible in beautiful print form. Sothis is beauty in ubiquity.

www.sothis.press
1 Waterhouse Square, London, ECIN 2ST

American Medical Botany

Being a Collection of the Native Medicinal Plants of the United States, Containing Their Botanical History and Chemical Analysis, and Properties and Uses in Medicine, Diet and the Arts

Jacob Bigelow

SOTHIS PRESS

ISBN 978-1-0365-0682-7

This edition is published by
Sothis Press © 2023. The company
is registered in England and Wales.
The Company Number is 14900389.

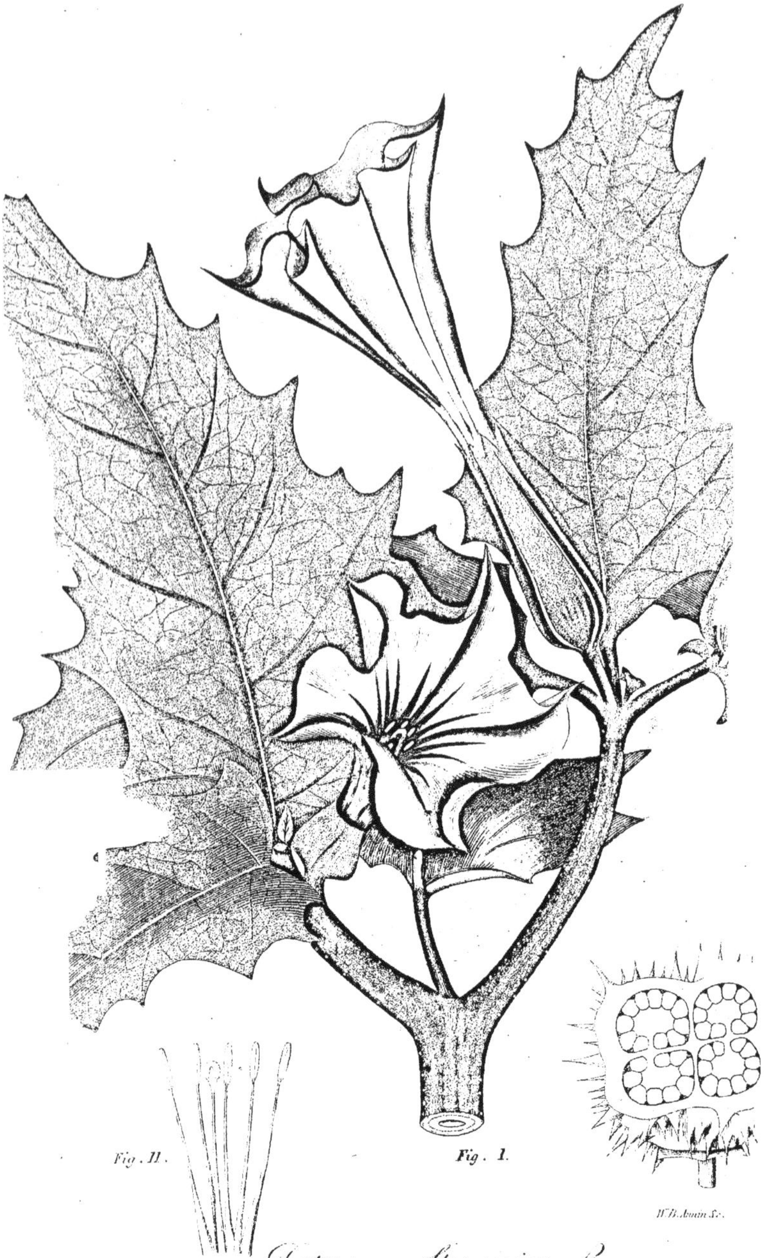

Fig. II.   Fig. 1.

W. B. Annin Sc.

Datura Stramonium β.

AMERICAN

MEDICAL BOTANY.

DATURA STRAMONIUM.

Thorn Apple.

PLATE I.

The Datura Stramonium is a wandering annual plant, which follows the progress of cultivation, and is rarely found remote from the vicinity of dwellings. It occurs in every part of the Atlantic coast from Maine to the Floridas, and is also found in the Western States in the neighbourhood of settlements. Its favorite haunts are the borders of fields and roadsides, among rubbish and in neglected spots of rich ground. It emigrates with great facility, and often springs up in the ballast of ships, and in earth carried from one country to another. This circumstance in Europe has undeservedly given rise to the opinion, that it is originally an American plant. Its native country, however, is doubtful, from the want of authentic descriptions of sufficient antiquity. One of the oldest satisfactory accounts of it is that of Gerarde in 1597, who has published a description and figure of this plant, and states that it was introduced into England by himself, from seeds received from Constantinople. [*Note* A.]

Its common name in Europe, derived from the form of its fruit, is *Thorn apple.* In this country its provincial names are *Apple* of *Peru*, *Devil's apple*, and *Jamestown weed.* It is a plant of rank growth and luxuriant foliage, varying in height from one to six feet, according to the soil in which it grows. In Carolina it begins to flower in May, and in Massachusetts about the latter part of July, and continues until the arrival of frosts.

The Datura Stramonium belongs to the first order of the fifth class in the Linnæan artificial arrangement. In its natural order it is found among the Luridæ of Linnæus and the Solaneæ of Jussieu. The following are the essential marks which characterize the genus Datura. *The corolla funnel form and plaited. The calyx tubular, angular and deciduous. The capsule four valved.*—Under this genus are comprehended a number of species, a great part of which are natives of warm latitudes. The species Stramonium is distinguished from the rest by the following character. *Capsules thorny, erect, ovate; leaves ovate, angular, smooth.*—A more particular description of the plant is as follows. Stem erect, simple at bottom, much branched at top by repeated forks, smooth or slightly pubescent, hollow in the large plants, often solid in small ones. Leaves given off from the forks of the stem, five or six inches long, acute, irregularly sinuated and toothed, with large acute teeth and round sinuses, the sides of the base extending unequally down the petiole. Flowers single, axillary, on short stalks, erect or nodding. Calyx composed of one leaf, tubular, with five angles and five teeth, deciduous by breaking off from its base. Corolla funnel shaped with a long tube, five angled, its margin waved and folded, and terminating in five acuminate teeth. Stamens growing to the tube by their filaments, with oblong erect anthers. Germ superior, hairy with the rudiments of spines, ovate; style as long as the stamens; stigma obtuse, parted at base. Capsule ovate, fleshy, covered with thorns, four valved, four celled, opening at top. Seeds numerous, reniform, black, attached to a longitudinal receptacle, which occupies the centre of each cell.

At least two distinct varieties of Datura Stramonium are common in the United States. One of these has a green stalk and white flowers, and agrees with the figures of Sowerby and Woodville, except that the anthers are somewhat longer and the dissepiment of the capsule thinner. The second variety, the one represented in our figure, has a dark reddish stem, minutely dotted with green; and purple flowers striped with deep purple inside. It is generally a larger plant, and its stem more universally hollow. This variety is probably the D. tatula of Linnæus, answering to the description in the Species plantarum. The distinguishing marks laid down between the two plants are not sufficient to make them distinct species. I have cultivated both together and watched them throughout their growth, without being able to detect any difference except in colour. Their sensible and medical properties are the same. Sir James Edward Smith has lately informed me, that on consulting the herbarium of Linnæus, the original specimens of D. Stramonium and tatula did not appear to be more than varieties of the same plant. [*Note* B.]

Every part of the Stramonium, when recent, has a strong, heavy, disagreeable odour, and a bitter, nauseous taste. Taken internally it proves a violent narcotic poison, affecting the mind and body in the most powerful manner. Its usual consequences when swallowed in considerable quantity, are vertigo and confusion of mind, insensibility of the retina, occasioning dilatation of the pupil and loss of sight, tremors of the limbs and loss of the power of voluntary motion, headach, dryness of the throat, nausea and vomiting, anxiety and faintness, and sometimes furious delirium. If the amount taken be large and not speedily ejected from the stomach, the symptoms pass into convulsions or lethargic stupor, which continue till death. When not fatal, its effects, like those of other narcotics, are temporary, disappearing in from one to two days, and frequently in a shorter period.—The remedies to be resorted to in cases of poison from Stramonium, are a prompt emetic, followed by a free use of vegetable acids and strong coffee.

Many stories have been related of the power of this and other species of Datura to produce mental alienation, without at the same time materially affecting the body. [*Note* C.] These accounts are generally of somewhat ancient date, and not correspondent with the observations of later physicians. They were suited to those days of credulity, in which the Royal Society of London gravely inquired of Sir Philberto Vernatti, "Whether the Indians can so prepare the stupifying herb Datura, that they make it lie several days, months, or years, according as they will have it, in a man's body; and at the end kill him without missing half an hour's time?"

Like opium and like other powerful medicines, this plant, when taken in small quantity, and under suitable regulations, proves a remedy of importance, and a useful agent in the hands of physicians. In common with some other narcotics, it seems first to have been introduced freely into practice by Baron Storck of Vienna, as a remedy in Mania, Epilepsy, Convulsions, &c. Many subsequent physicians have given testimony to its efficacy in certain forms of these disorders, yet the instances of its failure have doubtless been more frequent than those of its success. In Murray's Apparatus Medicaminum may be found a summary of the reports of many medical men, who have tried it with various success in the diseases in question, as well as in others. Dr. Cullen has no doubt that it may be a remedy in certain cases of mania and epilepsy; but doubts if any person has learned to distinguish the cases to which it is properly adapted.

Dr. Fisher, President of the Massachusetts Medical Society, has published in their communications some remarks on the employment of Stramonium in epilepsy. He divides the cases of that disease into three kinds; those of which the fits return daily; those in which they recur at regular periods, as monthly, or give warning of their approach by previous symptoms; lastly, those in which they do not observe any regular period, and do not give any warning of their approach. In the two first kinds he asserts, that all the cases which came under his care, and which were not very few, had been cured by Stramonium. In those of the third kind he found it of no benefit whatever.

Dr. Archer of Maryland has formed distinctions nearly similar in the application of Stramonium to epilepsy.

In a case of Tic doloureux of long standing I found the extract, taken in as large doses as the stomach would bear, to afford decided relief. Several practitioners have spoken to me of its efficacy in this formidable disease. It should be taken in large doses, and the system kept for some time under its influence.

Within a few years, the thorn apple has attracted much notice, both in Europe and in this country, as an efficacious palliative in Asthma and some other affections of the lungs, when used by smoking, in the same manner as tobacco.

The practice was first suggested by the employment of another species, the *Datura ferox*, for similar complaints, in the East Indies. An English gentleman, having exhausted the stock with which he had been supplied of the oriental plant, was advised by Dr. Sims to have recourse to the common Stramonium as a substitute; and upon trial, experienced the same benefit as he had done from the former species. This instance of success led to further trials, and in a short time several publications appeared, containing cases of great relief afforded by smoking this plant in the paroxysms of Asthma. Many individuals, of different ages, habits, and constitutions, had used it with the effect of producing immediate relief, and of terminating the paroxysm in a short time. The efficacy however of this medicine was called in question by Dr. Bree, a physician well known by his elaborate treatise on Asthma, who published in the Medical and Physical Journal a letter, containing the result of a great number of unsuccessful trials of Stramonium in asthmatic cases. It may be doubted whether any other physician has been so unfortunate in its use as Dr. Bree, since he affirms that not one case of those under his care was benefitted by it. Certain it is, that in this country the thorn apple is employed with very frequent success by asthmatic patients, and it would not be difficult to designate a dozen individuals in Boston and its vicinity, who are in the habit of employing it with unfailing relief in the paroxysms of this distressing complaint. The cases, which it is fitted to relieve, are those of pure spasmodic asthma, in which it doubtless acts by its sedative and antispasmodic effects. In those depending upon effusion of serum in the lungs, or upon the presence of exciting causes in the first passages, or elsewhere, requiring to be removed; it must not be expected that remedies of this class can afford benefit. In several cases of plethoric and intemperate people, I have found it fail altogether, and venesection afterwards to give speedy relief.

The part of the plant, which I have employed for smoking, is the leaf prepared in the same way as tobacco. The root, which has commonly been the part used, is more woody and fibrous, and possesses less of the juices of the plant, than its more pulpy and succulent parts. The root also, being strictly annual, has no opportunity to accumulate the virtues of the plant, beyond any other part.

In the seventh volume of the Medico-Chirurgical Transactions, for 1816, is a paper on the properties of the Stramonium by Dr. Marcet of London, Physician to Guy's Hospital. As the result of his experience, it appeared that this medicine taken internally had relieved acute pains of various kinds more effectually than any other narcotic substance. Its usual effects under his observation, when administered in appropriate doses, in chronic diseases attended with acute pain; were, to lessen powerfully and almost immediately sensibility and pain; to occasion a sort of nervous shock, which is frequently attended with a momentary affection of the head and eyes, with a degree of nausea, and with phenomena resembling those produced by intoxication; to excite in many instances nervous sensations, which are referred to the œsophagus or bronchiæ or fauces, and which sometimes amount to a sense like suffocation; to have rather a relaxing, than an astringent effect on the bowels; to have no marked influence on the pulse, except in a few instances to seem to render it slower; to produce but a transitory and inconsiderable dilatation of the pupil, and to have but little immediate tendency to produce sleep, except from the state of comparative serenity and ease, which follows the preceding symptoms.—In some instances its beneficial effects were obtained without the patient experiencing any of the uneasy sensations above mentioned.

The cases in which Dr. Marcet employed the Stramonium, with their results, appear in the following summary. In four cases of Sciatica, decided benefit was obtained. The efficacy of the medicine was still more strongly marked in two cases of sciatica combined with syphilitic pains. It failed in two instances of diseased hip joint. It produced considerable relief of pain in a case of supposed disease of the spine, followed by paraplegia; and likewise in one of cancer of the breast. It allayed materially the pain occasioned by an acute uterine disease. It was of great and repeated utility in a case of Tic doloureux, its utility in a second case of the same description was very doubtful, and in a third it entirely failed.

There are some authorities for the success of Stramonium in Chorea. Professor Chapman of Philadelphia has found it of use in dysmenorrhea, also with or without mercury in syphilitic and scrophulous ulcers of ill condition.

The external use of Stramonium is of much older date than its internal exhibition. Gerarde in his Herbal, published in 1597, says, "The iuyce of Thorne apples, boiled with hog's grease to the forme of an unguent or salve, cureth all inflammations whatsoever, all manner of burnings or scaldings, and that in very short time, as myself have found by my dayly practise, to my great credit and profit." Others, since the time of Gerarde, have used this preparation, if not with the same gratifying success, at least with some benefit as an anodyne, sedative application. It mitigates the pain in burns and inflammatory tumors, and promotes the cure of certain cutaneous eruptions. In some irritable ulcers with thickened edges and a sanious discharge, I have found it remarkably efficacious in changing the condition and promoting the granulations and cicatrization. In painful hemorrhoidal tumors the ointment of Stramonium with the ointment of acetate of lead gives, in many cases, very prompt and satisfactory relief, being in this respect inferior to no application, with which I have been acquainted.

Applied topically to the eye, the preparations of Stramonium diminish the sensibility of the retina, and relax the iris. From this effect it is employed by many surgeons to dilate the pupil, as preparatory to the operation for cataract.

The virtues of Stramonium appear to be seated in an extractive principle, which dissolves in water and alcohol, but most readily in the former. It is copiously precipitated from the infusion by muriate of tin. With sulphate of iron it gives a deep green colour, and with gelatin suffers no change. Water distilled from the plant has the sensible qualities in a slight degree, but does not seem to possess the medicinal powers of the plant. Dr. S. Cooper, in a valuable dissertation on this plant, says, that an ounce of the distilled water was taken into the stomach with little or no effect. The same gentleman states, that upon evaporating the infusion of Stramonium, he observed a large number of minute crystals, resembling particles of nitre. Thinking it possible that these might be something analogous to the crystals, said to be obtained by Derosne from opium, and by him denominated the narcotic principle, I repeated the experiment by carefully evaporating separate decoctions of the green and dried leaves. No crystals however were discoverable at any stage of the process, either to the touch, or to the eye assisted by a strong magnifier.

The forms in which the Stramonium is prepared for use are the powder, the inspissated juice, the extract, the tincture and the ointment. The *powder* should be made as soon as the plant is dry, and kept in close stopped bottles.—The *inspissated juice* is made by compressing the bruised leaves in a strong bag, until the juice is forced out. This is to be evaporated in flat vessels at the heat of boiling salt water to the thickness of honey; it is then suffered to cool, put up in glazed vessels and moistened with alcohol. The *extract* is prepared by immersing a pound of the leaves in three gallons of water and boiling down to one. The decoction should then be strained and stand six hours to settle, after which it may be drawn off and evaporated to the proper consistence. When the seeds are used, the decoction should stand a longer time to separate the oil with which the cotyledons abound, before evaporation. A larger amount of extract may be obtained by boiling the portion, which has been used, a second time in a smaller quantity of water, and mixing the two decoctions before evaporation. For the *tincture* one ounce of the dried leaves is to be digested for a week in eight ounces of proof spirit, and filtrated through paper. In making the *ointment*, a pound of the fresh leaves may be simmered in three pounds of hog's lard until the leaves become crisp. It is then to be strained, and cooled gradually.

The period for gathering the leaves is from the time the plant begins to flower, until the arrival of frost.

As the preparations of Stramonium are liable to vary in strength according to the circumstances under which they are made, it is always prudent to begin with the smallest dose, and repeat it about three times a day, increasing each dose until the effects begin to appear in the stomach or head.

The commencing doses of the Stramonium, when properly prepared, are as follows.

Of the powdered leaves	*1 grain.*
powdered seeds	*½ a grain.*
inspissated juice or extract	*1 grain.*
extract of the seeds	*from ⅛ to ¼ grain.*
tincture	*from 15 to 20 drops.*

BOTANICAL REFERENCES.

Datura Stramonium, LINNÆUS *Sp. pl. Fl. Suec.* 185 *&c.*—GRONOVIUS *Fl. Virg.* 23.—ŒDER. *Fl. Danica* 436.—BLACKWELL *t.* 313.—GMELIN *Iter* i. 43.—POLLICH. *Palatin.* 224.—HOFFMANN *Germ.* 77.—ROTH *Fl. Germ.* i. 92 *&c.*—WOODVILLE *t.* 124.—CURTIS *Lond.* vi. *t.* 17.—SMITH *Fl. Brit.* 254.—*Engl. Bot. t.* 1288.—PURSH *Amer.* 141.—ELLIOTT *Carol.* i. 275.—Stramonium foliis angulosis &c. HALLER *Helv.* 586. Nuci metellæ congener planta, CAMERARIUS *Epitome* 276.—Solanum fœtida pomo spinoso, oblongo, &c. BAUHIN *pin.* 168.—Stramonium spinosum, GERARDE *Herbal* 348.

MEDICAL REFERENCES.

STORCK *de Stramonio &c.*—LINDENSTOLPE *de venenis,* 531.—SAUVAGES *Nosol.* 2. 430.—GREDING in *Ludwigs Adversaria* i. 345.—MURRAY *App. Med.* i. 670.—CULLEN *Mat. Med.* ii. 281.—FOWLER *in Med. Comment.* v. 161.—ODHELIUS *cit. in Med. Comment* v. 161.—PAPIN *in Phil. Trans.* abr. vi. 53.—RUSH *in Philad.*

Trans. i. 384.—Schoepf. 24.—WEDENBERG *in Med. Comment* iii. 18.—BEVERLY, *Hist. Virg.* p. 121.—Medical and Physical Journal, vol. xxv. & xxvi. *in various places.* COOPER *in Caldwell's Theses,* vol. i.—BARTON, *Coll. Mat. Med.* 46.—CHAPMAN *in edit.* MURRAY 146.—THATCHER, *Disp.* 205.—MARCET *Medico-Chirur. Trans.* vii.

PLATE I.

Fig. 1. *A branch of Datura Stramonium, the purple variety, with leaves and flowers.*

Fig. 2. *Stamens and style.*

Fig. 3. *Transverse section of the pericarp, shewing the cells, receptacles and seeds.*

Eupatorium perfoliatum

EUPATORIUM PERFOLIATUM.

Thorough wort.

PLATE II.

THE peculiar form and arrangement of the leaves in this plant render it very easy of distinction at sight by the most inexperienced botanist. It flowers from midsummer to September, and is found in all latitudes from Nova Scotia to Florida. It inhabits meadows and boggy soils, growing most frequently in bunches, the stems being connected by horizontal roots. Its common names are *Thorough wort, Thorough wax, Cross wort, Bone set, &c.*

The genus Eupatorium, belonging to the first order of the class Syngenesia or Compound flowers, and to the order Corymbiferæ of Jussieu, is characterized by its *naked receptacle, its down simple or rough, its calyx oblong and imbricate, its style longer than the corolla, and cloven half way down.* The species perfoliatum, exclusively an inhabitant of America, is abundantly distinguished from the rest, by the peculiar form of its leaves, indicated in its name. Michaux has altered the specific name to *connatum* I think injudiciously.

The stems of this plant are erect, round, hairy branched at top only. The leaves, which are perforated by the stem, are rather perfoliate than connate, since they have not the character of two leaves joined together, but of one entire leaf, having its four principal veins proceeding at right angles from the four quarters of the stem, two of them being situated in the place of the supposed junction. The upper leaves however are generally divided into pairs. The main leaves are acuminate, decreasing gradually in breadth from the stem, where they are widest, to the extremities. They are serrated, wrinkled, pale underneath, and hairy, especially on the veins. Flowers in corymbs with hairy peduncles. Calyx cylindrical, imbricate, the scales lanceolate, acute, hairy. Each calyx contains about twelve or fifteen florets, which are tubular, with fine spreading segments, and surrounded with a rough down. The stamens in each consist of five soft filaments, with blackish anthers united with a tube. Style filiform, divided into two branches, which project above the flower. Seeds oblong on a naked receptacle.

Every part of the Eupatorium has an intensely bitter taste, combined with a flavour peculiar to the plant, but without astringency or acrimony. The leaves and flowers abound in a bitter extractive matter, in which the important qualities of the plant seem to reside. I find this principle to be alike soluble in water and alcohol, imparting its sensible qualities to both, and neither solution being rendered turbid, at least for some time, by the addition of the other solvent. It forms copious precipitates with many of the metallic salts, such as muriate of tin, nitrate of mercury, nitrate of silver, and acetate of lead. Of the mineral acids, the sulphuric and muriatic form slight precipitates with the aqueous decoction; the oxymuriatic, a more copious one; the nitric, in my experiments, gave no precipitate, but changed the colour to a red. In the alcoholic solution the oxymuriatic alone formed an immediate precipitate. Tannin exists very sparingly in this plant. A solution of isinglass produced a slight precipitate from the tincture, and a hardly perceptible turbidness in separate decoctions of the leaves and flowers. Sulphate of iron gave a dark green precipitate, which partially subsided in a short time.—In distillation, water came over very slightly affected with the sensible qualities of the plant, and not alterable by sulphate of iron.

A dissertation of merit on this plant was published a few years since by Dr. Anderson of New York, in which he gives the details of numerous and elaborate chemical trials, made by him on different parts of the plant. He concludes, among other things, from his experiments, that the active properties of the plant reside in greatest quantity in the leaves, and that its virtues are readily obtained by means of a simple decoction.

The medical powers of Eupatorium are such as its sensible properties would seem to indicate, those of a tonic stimulant. Given in moderate quantities, either in substance or in cold infusion or decoction, it promotes digestion, strengthens the viscera, and restores tone to the system. Like other vegetable bitters, if given in large quantities, especially in warm infusion or decoction, it proves emetic, sudorific, and aperient. Even in cold infusion it tends to bring on diaphoresis.

This plant has been long in use in different parts of the United States, for the same purposes for which the Peruvian bark, Gentian, Chamomile, &c. are employed. It has been found competent to the cure of intermittent fevers by various practitioners in the middle and southern states. Dr. Anderson has detailed six cases of intermittents, quotidian, tertian, and quartan, out of a large number which had been successfully treated within his own observation by the Eupatorium both in substance and decoction. In these cases the cures were certainly expeditious, and took place at as early a period as could have been expected from arsenic or the Peruvian bark. Dr. A. cites the experience of several distinguished practitioners, particularly Dr. Hosack of New York and the late Dr. Barton of Philadelphia, in confirmation of his own, to shew that the Eupatorium is an efficacious remedy in the treatment of various febrile disorders, also of many cutaneous affections, and diseases of general debility.

I have prescribed an infusion of the Eupatorium in various instances to patients in the low stages of fever, where it has appeared instrumental in supporting the strength and promoting a moisture of the skin, without materially increasing the heat of the body. I have also found the cold infusion or decoction a serviceable tonic in loss of appetite and other symptoms of dyspepsia, as well as in general debility of the system.

The warm infusion is a convenient substitute for that of chamomile flowers in facilitating the operation of an emetic.

When employed as a tonic, this plant may be taken in powder in doses of twenty or thirty grains, or a teacup full may be used of the infusion, rendered moderately bitter. When intended to act as an emetic, a strong decoction may be made from an ounce of the plant in a quart of water, boiled to a pint.

BOTANICAL REFERENCES.

Eupatorium perfoliatum, LINNÆUS, *Sp. pl.*—AITON, *Hort. Kew.* iii. 160.—WILLDENOW, *Sp. pl.* iii. 1761.—GRONOVIUS, *Virg.* 119.—COLDEN, *Novebor.* 181.—STOKES, iv. 171.—PURSH, ii. 516.—Eupatorium connatum, MICHAUX, *Fl. Amer.* ii. 99.—Eupatorium Virginianum, &c.—PLUKENET, *t.* 87. *f.* 6.—RAIUS, *suppl.* 189.—MORISON, *hist.* iii. 97.

MEDICAL REFERENCES.

SCHOEPF 121.—GUTHRIE *in Annal. Med.* iii. 403.—BART. *Coll.* 28.—*Med. and Phys. Journal.*—THACHER *Disp.* 217.—ANDERSON, *Inaugural Thesis.*

PLATE II.

Fig. 1. *Eupatorium perfoliatum.*

Fig. 2. *A flower magnified.*

Fig. 3. *A floret magnified.*

Fig. 4. *Tube of anthers with the style running through.*

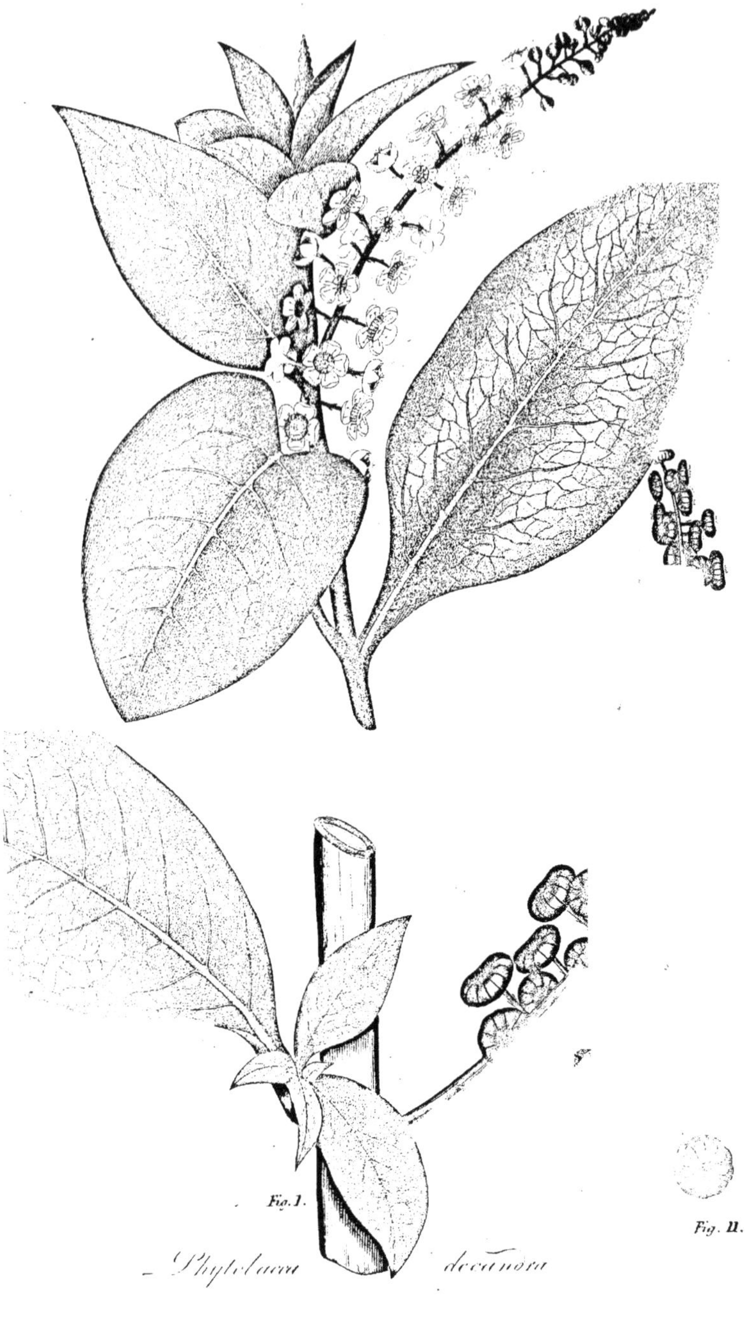

Fig. I.

Fig. II.

Phytolacca decandra

PHYTOLACCA DECANDRA.

Poke.

PLATE III.

From the testimony of different writers it appears, that the Phytolacca decandra is an inhabitant not only of North America, but likewise of the south of Europe from Portugal to Greece, and also of the Barbary states in Africa. Its origin is probably American, since I find that it was so considered in the time of Parkinson, who in his Theatrum Botanicum, published in 1640, denominates it "Solanum magnum Virginianum rubrum." This is one of the oldest accounts I find of it. Plukenet conjectures it may be the *Cuechiliz tomatl* of Hernandez, but the description, like most others of that loose and superficial writer, are more promotive of obscurity than of knowledge, and it is not easy to draw from it any satisfactory evidence as to its Mexican origin. [*Note D.*]

In the autumnal months no plant among us is more remarkable than the Phytolacca for its large size, and the fine colour of its clusters of berries. Its most general appellation is *Poke*, an abbreviation, perhaps, of *Pocan*, the name by which it was known in Virginia a century ago. In New England it is more frequently called *Garget*, *Cocum*, *Jalap* and *Pigeon berries*.

Jussieu has arranged this genus among his *Atriplices*, and Linnæus with the *Oleraceæ*.

The number of its stems and styles, place it in the class *Decandria* and order *Decagynia*. Its generic character consists in having *no calyx, a corolla of five petals, and its berries superior with ten cells and ten seeds.* The species *decandra* is the only one which strictly agrees with its class and order, and is known by having *ovate leaves, acute at both ends, and its flowers with ten stamens and styles.*

The root of this plant is of large size, frequently exceeding a man's leg in thickness, and is usually divided into two or three principal branches. Its substance is fleshy and fibrous, and easily cut or broken. Internally it is distinctly marked with concentric rings of considerable thickness, while its outer surface is covered with a very thin brownish bark, which seems to be little more than a cuticle. The stalks, which are annual, frequently grow to the height of six, and even nine feet. They are round, smooth, and very much branched. When young, their usual colour is green, but in most plants, after the berries have ripened, they are of a fine purple. The leaves are scattered, petioled, ovate-oblong, smooth on both sides, ribbed underneath, entire, acute. The flowers grow on long pedunculated racemes opposite to leaves. Peduncles nearly smooth, angular, ascending. Pedicels divaricated, sometimes branched, green, white, or purple, furnished with a small linear bracte at base, and two others in the middle. Calyx none. Corolla resembling a calyx, whitish, consisting of five round-ovate, concave, incurving petals. Stamens ten, rather shorter than the petals, with white, roundish, two lobed anthers. Germ greenish, round, depressed, ten furrowed. Styles ten, short, recurved. The flowers are succeeded by long clusters of dark purple berries, almost black, depressed or flattened, and marked with ten furrows on the sides.

The dried root is light coloured and spongy, with a mild and somewhat sweetish taste. A part of it is soluble both in water and alcohol, and neither of these substances renders turbid the solution in the other, unless the solution has been inspissated by long boiling. The soluble portion appears neither resinous nor mucous. It approaches most nearly to extractive, but has characters somewhat peculiar to itself. A decoction of the root procured by boiling for ten minutes in distilled water, exhibited after filtration the following results. It was transparent, nearly colourless, and did not alter litmus. It gave no precipitate with the sulphuric, nitric, muriatic, oxymuriatic, and acetous acids. It gave no precipitate with the sulphate of iron, but formed a copious one with the nitrates of mercury and silver, and the acetate of lead. Muriate of tin produced no effect at first, but after standing, a light precipitate took place. Pearl ash, lime water, and muriate of barytes rendered the solution turbid. Acetate of barytes occasioned no change. Oxymuriate of lime formed an immediate precipitate.

The cold infusion exhibited nearly the same results as the decoction. The alcoholic solution underwent no change from muriate of tin, but threw down a dense precipitate with nitrate of mercury.

From the above experiments it appears, that the soluble principle of the Phytolacca differs from common vegetable extractive, as defined by the chemists, in several respects, particularly in not being thrown down by the oxymuriatic or other mineral acids, and in being but partially affected by muriate of tin.

In the *Annales de Chimie*, vol. lxxii, is a memoir on the Chemical properties of the Phytolacca decandra by M. Braconnot. His experiments indicate the presence of an unusual quantity of vegetable alkali in this plant. He found that the ashes, procured by incinerating the stalks, afforded nearly 67 per cent. of dried alkaline carbonate, and 42 per cent. of pure caustic potash. This alkali in the plant is neutralized by an acid having considerable affinity to the malic, but with a few shades of difference. With lime and lead malic acid forms flocculent precipitates, very easily soluble in distilled vinegar, but those with the phytolaccic acid are insoluble. M. Braconnot thinks this acid may probably be a mean between the malic and oxalic acids, or an oxygenized malic acid.

The same memoir contains an examination of the colouring matter in the berries of the Phytolacca. The juice of these berries is of a very fine, bright purple colour, but this colour is extremely fugacious and disappears in a short time from cloth or paper that has been tinged with it. A few drops of lime water added to this purple juice change it to a yellow colour, but the smallest quantity of acid is sufficient to restore its purple hue. Exposure to the air or large dilutions is sufficient to restore the original purple.

M. Braconnot considers the yellow liquor produced by the juice of these berries and lime water as one of the most delicate tests of the presence of acid. Into two glasses he put equal quantities of the juice made yellow and of an infusion of litmus of equal depth of colour. More than sixty drops of a very weak acid were required to redden the infusion of litmus, but less than fifteen restored the purple colour of the Phytolacca. Hence it follows, that the yellow liquor is four times as sensible to the presence of acid, as the infusion of litmus. It however requires to be used immediately after it is prepared, since a few hours cause a spontaneous change in it, which begins with a precipitate, and ends with a deprivation of colour.

The effects produced on this purple colour by other reagents were as follows. Pure alkalis gave it a yellow colour. Alkaline subcarbonates a violet, that fades and becomes yellow by standing. Weak acids no perceptible change. Dilute oxymuriatic acid a complete deprivation of colour with white flocculi. Alum nothing at first, but after some days, a very light red precipitate. Muriate of lime no change. Muriate of tin a red sediment inclining to lilac, leaving the fluid colourless. Nitrate of lead a precipitate of the colour of wine lees. Super oxided sulphate of iron, a dirty violet.

Many of the above experiments I have repeated, and added others. The yellow colour produced by the alkalis borders on green. Pure strontian produces the same change as potash and lime. Pure barytes wholly discharges the colour on standing a short time. Acetate of lead forms a scarlet precipitate, leaving the liquid nearly colourless.

The purple colour that tinges the cuticle of the stalks of the Phytolacca is stated in the above memoir, to be of the same nature as that in the berries, and to afford the same results.

The taste of the berries is sweetish and nauseous, leaving behind a very slight sense of acrimony. M. Braconnot, found that at a moderate temperature, the juice underwent the vinous fermentation, and yielded alcohol by distillation. Dr. Shultz procured from half a bushel of the berries six pints of spirit sufficiently strong to take fire and burn with readiness.

In its medicinal properties the root of the Phytolacca decandra approaches nearer to ipecacuanha than any American vegetable, I have hitherto examined. From abundant experience, the result of many trials made in Dispensary practice, I am satisfied that, when properly prepared, it operates in the same doses and with the same certainty, as the South American emetic. Ten grains of the powder will rarely remain on the stomach, and twenty or thirty produce a powerful operation, by emesis and generally by catharsis. In its mode of operation, this medicine has some peculiarities, a part of which are favorable, others disadvantageous. Its advantages are, that it operates with ease, and seldom occasions pain or cramp. Its disadvantages are, 1. That it is slow in its effects, frequently not beginning to operate until an hour, and sometimes two hours after it is taken. 2. That it continues to operate for a greater length of time than is usual for emetics, although as far as I have been able to observe, it is readily checked by an opiate. These disadvantages however are not constant. I have repeatedly known it commence operating in fifteen minutes, and cease after four or five ejections. The representations of patients as to any unpleasant feelings under its effects, are not greater than we should naturally expect, when it is recollected, that no emetic is altogether comfortable in its operation. Dr. Fisher of Beverly* informs me that whenever he has used the Phytolacca, it has performed its duty as an emetic perfectly well, and that in one patient, a female of irritable stomach, in whom previous emetics had always excited severe spasms, ten grains of the Phytolacca operated effectually, and no spasm followed.

I have sometimes observed slight narcotic symptoms during the operation of Phytolacca, particularly vertigo. But others have not always met with this symptom. Dr. George Hayward of this town, who has had much experience with this medicine, the results of which were communicated to the Linnæan society, and afterwards published in the New England Journal, October 1817, states that in doses of a scruple, he never noticed any dizziness, or stupor from it, although he had always been particular in his inquiries to know if any such symptoms took place. The above dose was administered by him in nearly thirty cases, in all of which, except in one case, it operated as an emetic and cathartic, usually three or four times, thoroughly, though not severely, generally commencing its operation on the stom-

* Letter dated November, 1815.

ach in an hour, and rarely continuing longer than four. He found it to excite little or no nausea previous to its operation, and though it made a powerful impression on the system, it never produced any disagreeable or unusual symptoms.

Dr. Hayward also made trial of the powder of the leaves, which he found to possess the same properties with that of the root, but to be less effectual and less certain in its operation. He also prepared a tincture, decoction, and wine of the root; but all these were inferior to the medicine in substance, being less certain in their effect, and sometimes giving rise to troublesome symptoms.

Dr. Shultz of Pennsylvania, author of an inaugural dissertation on the Phytolacca decandra, gave the expressed juice of the leaves, berries, and roots, in considerable quantity to animals. It operated by emesis and catharsis, attended with drowsiness. The juice of the root was most active. He also gave to a dog two ounces of the spiritous liquor distilled from the berries. It occasioned nausea and drowsiness, with slight spasmodic motions, but no vomiting.

In the same dissertation, Dr. Shultz refers to several instances of persons who had incautiously eaten large quantities of the root through mistake. Its effects were violent vomiting and purging, prostration of strength, and in some instances convulsions.

The Phytolacca has had some reputation in the treatment of rheumatism. Dr. Griffits, formerly a professor in the University of Pennsylvania, found it of great use in Syphilitic rheumatism. Dr. Hayward however states, that he derived no advantage from its employment in rheumatic affections.

The young shoots of this vegetable are destitute of medicinal qualities, and are eaten in the spring in some parts of the United States, as substitutes for asparagus. At this time the *succus proprius* or returning juice of the plant is not yet formed by exposure of the sap to the atmospheric air, in the leaves. The ripe berries are less noxious than the green, and are devoured by several species of birds. In Portugal and in France they were formerly employed to improve the colour of red wines, until the interference of government became necessary to put a stop to the practice.

The external application of Phytolacca has been found useful in a variety of cases, by its action as a local stimulant. The ointment and extract have commonly been employed for this purpose. These preparations usually excite a sense of

7

heat and smarting on being first applied. I have cured cases of psora with the ointment, and Dr. Hayward states, that he found it successful in cases where sulphur had failed. A case of tinia capitis of twelve years' standing, which had resisted various kinds of treatment, was also cured by this application.

The Phytolacca is one of those vegetables which has had its temporary reputation for the cure of cancer. For this purpose it has been resorted to in various parts of the world, and many men of science have been convicts to its efficacy, among whom were Dr. Colden and Dr. Franklin of our country. [*Note* E.] But like other vegetable specifics for cancer, it owes its character to an imperfect discrimination of that disease, and a misapplication of the name. All that can be strictly inferred from the various accounts we have had on this subject, is, that the plant has often proved useful in malignant ulcers by its stimulating and almost escharotic effects, frequently producing an eschar, and thus altering the condition of the ulcerated surface.

For internal use no preparation of the Phytolacca is to be preferred to the powder, of which from ten to fifteen grains is often a sufficient emetic.

The root should be dug late in autumn or during the winter. It should be cut in transverse slices and dried. After being pulverized, it should be kept in close stopped phials. The stock should be annually renewed, as its activity is impaired by age.

BOTANICAL REFERENCES.

Phytolacca decandra, LINNÆUS, *sp. pl.*—AITON, *Hort. Kew.* ii. 122.—*Botanical Magazine, t.* 931.—MICHAUX, *Fl. Amer.* i. 278. PURSH. i. 324.—Phytolacca vulgaris, DILLENIUS, *Hort. Elth. t.* 239.—P. Americana—BOERHAAVE, *Hort. Lug.* ii. 70.—Solanum racemosum Americanum, RAIUS, *Hist.* 662.—PLUKENET, *Phyt. t.* 225. *f.* 3.—Solanum magnum Virginianum rubrum, PARKINSON, *Theatrum,* 347.—Blitum Americanum, MUNTINGIUS, *Phyt. cur. t.* 212.

MEDICAL REFERENCES.

MURRAY, *appar. med.* iv. 335.—KALM, *travels in N. Amer.* i. 197.—GRAFFENREID, *Mem. Berne,* iii. 185.—SCHŒPF. 71.—BROWNE, *Hist. Jamaica,* 232.—*Amœn. Acad.* iv.—MILLER, *Dict. under the name.*—SPROGEL. *Diss. cir. ven.* 24.—BECKMAN, *comment. Gotting.* 1779, 74.—ALLIONI, *Flor. Ped.* ii. 132.—FRANKLIN, *works, vol.* i.—CUTLER, *Mem. Amer. Acad.* i. 447.—RUSH, i. 259.—THACHER, *Disp.* 300.—SHULTZ, *Inaugural thesis.*—HAYWARD, *N. Engl. Journal,* vi.

PLATE III.

Fig. 1. *Phytolacca decandra in flower and in fruit.*

Fig. 2. *Section of a berry.*

ARUM TRIPHYLLUM.

Dragon root.

PLATE IV.

It appears, that both North and South America give rise to this species of Arum, which is so versatile in its constitution as to bear the winters of Canada, and the perpetual summer of Brazil. In its structure it is one of our most singular vegetables, and in colour one of the most variable. It grows in swamps and damp shady woods, and is universally known among us by the names of *Dragon root* and *Indian turnip.*

The class to which the family of Arums belong, is rendered somewhat obscure by the variation of the species. Most botanists have placed them in the class *Monœcia*, others in *Polyandria.* The species under consideration is undoubtedly *Polygamous.* In natural arrangements, the Arums

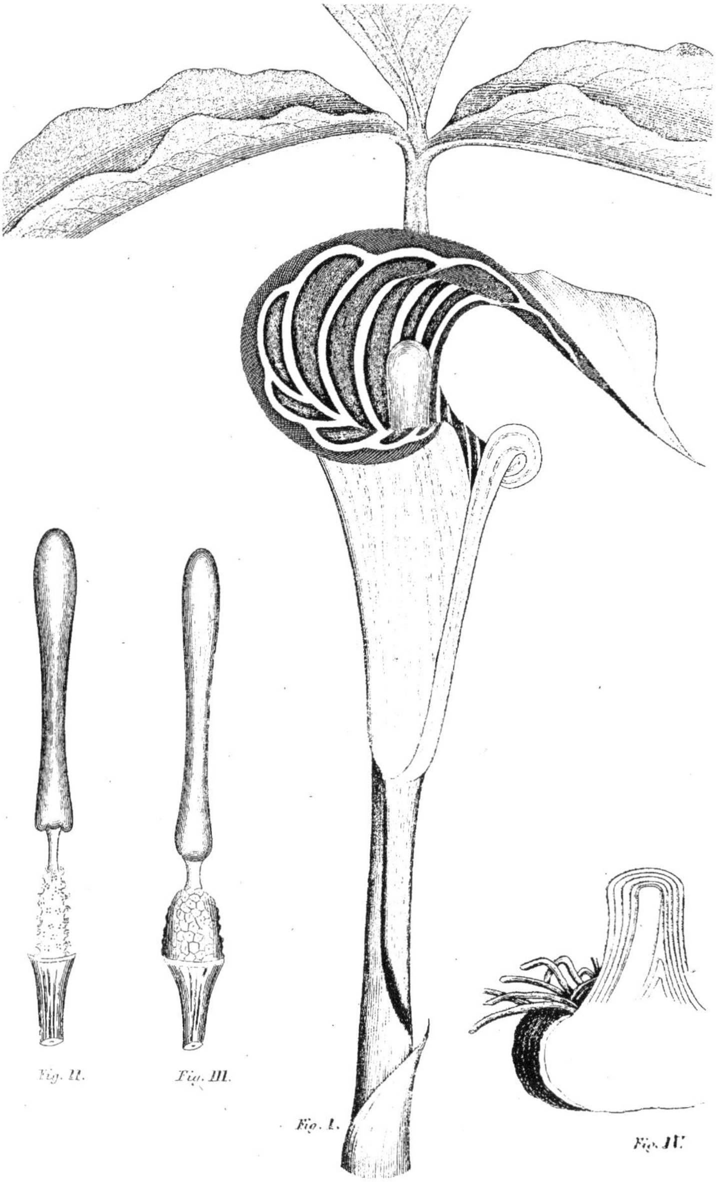

Arum triphyllum

gether. In some it is wholly green, in others dark purple or black. In most it is variegated, as in our figure, with pale greenish stripes on a dark ground. The spadix is much shorter than the spathe, club shaped, rounded at the end, green, purple, black, or variegated, suddenly contracted into a narrow neck at base, and surrounded below by the stamens or germs. In the barren plants, its base is covered with conical, fleshy filaments, bearing from two to four circular anthers each. In the fertile plants, it is invested with roundish crowded germs, each tipt with a stigma. Plants which are perfectly monœcious, and which are the least common, have stamens below the germs. There are also frequently found irregular, reniform substances, much larger than the anthers, of which they seem to be a disease. The upper part of the spadix withers with the spathe, while the germs grow into a large compact bunch of shining scarlet berries.

Every part of the Arum, and especially the root, is violently acrid, and almost caustic. Applied to the tongue or to any secreting surface, it produces an effect like that of Cayenne pepper, but far more powerful, so much so, as to leave a permanent soreness of many hours' continuance. Of this any one may become satisfied by a simple application of the root to his mouth. Its action does not readily extend through the cuticle, since the bruised root may be worn upon the external skin until it becomes dry, without occasioning pain or rubefaction.

The acrid property, which resides in this and other species of Arum, appears to depend upon a distinct vegetable principle in Chemistry, at present but little understood. It is extremely volatile, and disappears almost entirely by heat, drying, or simple exposure to the air. I have endeavoured, with but partial success, to obtain it in a separate state, or in any perceptible combination. The following were some of the methods by which it was attempted.

Portions of the fresh contused root were separately digested in water, in proof spirit, in alcohol, in ether, in olive oil and in vinegar. The infusions were tasted at different periods, but none of them had acquired the least acrimony from the plant.

The expressed juice of the root upon standing one minute had lost all its pungency.

A quantity of the bruised root was placed in a retort and covered with water. Heat was gradually applied, until a fluid began to collect in the receiver. This fluid had the peculiar odour of the root, but was wholly without acrimony. The same experiment was repeated with alcohol, and vinegar, and afforded similar results. In every case the liquid remaining in the retort was also without pungency.

Some slices of the root were digested in proof spirit in a close stopped phial. The portions of root retained their acrimony at the end of some weeks, but had imparted none to the spirit. At the end of two years, the root was examined and found destitute of acrimony, as were also the whole contents of the phial.

Suspecting that the acrid principle of this plant must escape in form of gas during the processes which have been mentioned, the following experiment was made. A quantity of the bruised root and stalks were placed in a vessel of water. A glass receiver was filled with water and inverted over them, and sufficient heat applied to raise the water nearly to the boiling point. From the beginning of the process, bubbles of air continued to escape from the plant, and were collected in the upper part of the receiver. In the course of half an hour, a considerable quantity of permanent gas was obtained. A part of this gas, after cooling, was transferred to a phial, in which was a small quantity of atmospheric air. On presenting a lighted paper to the mouth of this phial, it exploded with a very distinct report. Another portion of the gas was agitated with lime water, which it rendered turbid. This circumstance was probably owing to the mixture of carbonic acid disengaged from the plant, or from the water by boiling.

From the above experiments, which circumstances did not permit me to pursue, it appears that the acrimony of the Arum resides in a principle having no affinity for water, alcohol, or oil, being highly volatile, and, in a state of gas, inflammable. The products of its combustion, as well as its other affinities, remain to be investigated.*

The acrimony of the Arum when fresh is too powerful to render its internal exhibition safe. The roots, when dried whole, retain a small portion of their pungency, and in this state they have been given by some practitioners in the country for flatulence, cramp in the stomach, &c. also for

* The acrimony of the Ranunculi, which approaches that of the Arum, is lost by drying, yet is soluble in water, and passes over with it in distillation. That of Polygonum hydropiper disappears in decoction and distillation. The same takes place with several other acrid plants which I have examined. Some inquiries into the *acrid principle* of vegetables I am in hopes to render more mature at a future period.

asthmatic affections. As topical stimulants, they promise to be useful when any method shall have been discovered of fixing and preserving their acrimony. The late Dr. Barton of Philadelphia observes, that "the recent root of this plant boiled in milk, so as to communicate to the milk a strong impregnation of the peculiar acrimony of the plant, has been advantageously employed in cases of consumption of the lungs." This statement however should be qualified by the recollection, that the Arum imparts none of its acrimony to milk upon boiling. An impression of this kind can only have been received from a partial mixture of the substance of the root with the milk.

The root contains a large proportion of very pure white fæcula, resembling the finest arrow root or starch. To procure this, the fresh root should be reduced to a pulp, and placed on a strainer. Repeated portions of cold water should then be poured on it, which in passing through the strainer carry with them the farinaceous part, leaving the fibrous portion behind. The fæcula thus obtained, loses its acrimony on being thoroughly dried, and forms a very white, delicate and nutritive substance. Dr. M'Call of Georgia found these roots to yield one fourth part of their weight of pure amylaceous matter.—It is not uncommon for a nutritious fæcula to exist in pungent and poisonous roots. The Laplanders prepare a wholesome bread from the acrid roots of *Calla palustris*, and the juice of the *Cassava*, or bread root tree of the West Indies, is known to be highly deleterious while recent. [*Note* F.]

BOTANICAL REFERENCES.

LINNÆUS, *sp. pl.*—WILLDENOW, iv. 480.—AITON, *Hort. Kew.* iii. 315.—WALTER, *Carol.* 224.—MICHAUX, *Fl.* ii. 188.—PURSH, ii. 399. Dracunculus s. Serpentaria triphylla, &c.—BAUHIN, *Pin.* 195.—Arum s. Arisarum, &c.—MORISON, *Hist.* iii. 547, s. 13, *t.* 5.—PLUKENET, *t.* 77, *f.* 5. also *t.* 376, *f.* 3.

MEDICAL REFERENCES.

SCHŒPF, *Mat. Med.* 133.—RUSH, ii. 301.—BARTON, *Coll.* 29, &c.—M'CALL, *in Philad. Med. and Phys. Journal*, ii. 84.—THACHER, *Disp.* 153.—CUTLER, *Mem. Amer. Acad.* i. 487.

PLATE IV.

Fig. 1. *Arum triphyllum.*

Fig. 2. *Spadix with anthers.*

Fig. 3. *Spadix with germs.*

Fig. 4. *Longitudinal section of the root.*

COPTIS TRIFOLIA.

Gold thread.

PLATE V.

THE dark sphagnous swamps, which in the northern parts of our continent are covered with a perpetual shade of firs, cedars and pines, are the favourite haunts of this elegant little evergreen. The coldest situations seem to favour its growth, and it flourishes alike in the morasses of Canada and of Siberia. On our highest mountain tops it plants itself in little bogs and watery clefts of rocks, and perfects its fructification in the short summer allowed it in those situations. I have gathered it upon the summit of the Ascutney in Vermont, and on the Alpine regions of the White mountains. It is here that in company with the Diapensia and Azaleas of Lapland, the blue Menziesia, the fragrant Alpine Holcus, and other plants

Pl. V.

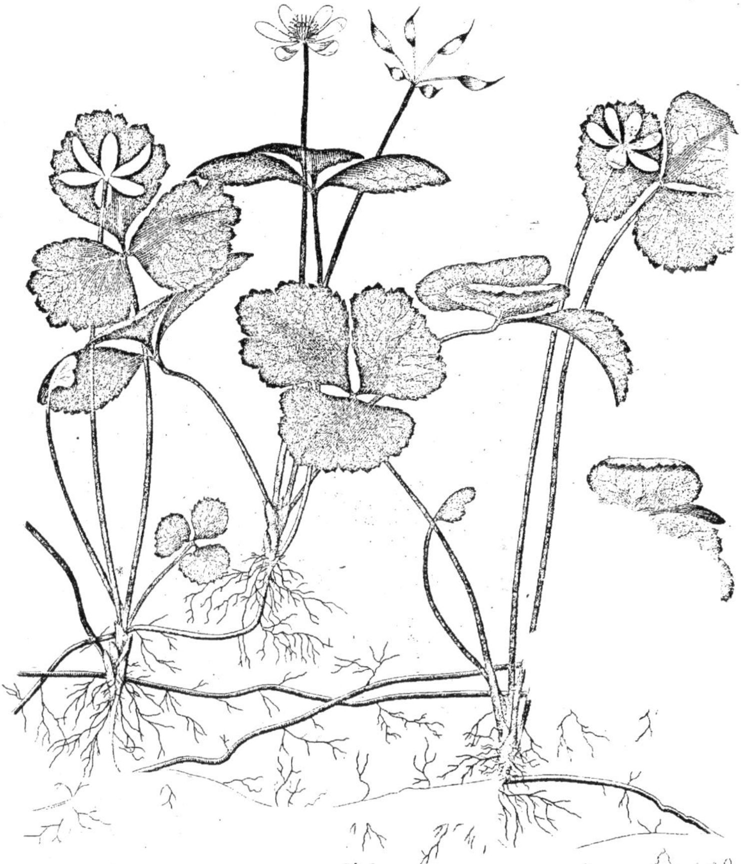

Fig. I.

Fig. II.

Fig. III.

Coptis trifolia

of high northern latitudes, it forms the link of botanical connexion between the two continents. When in situations like this, we seem transported to the frigid zone, and to be present at the point where the hemispheres approach each other, as if to interchange their productions.*

In the second volume of the Amœnitates Academicæ is a description and imperfect figure of this plant as brought from Kamschatka, by Halenius. He describes it by the name *Helleborus trifolius*, with the observation, "Minima est hæc planta in suo genere, attamen spectabilis." Subsequent botanists have ranked it with the Hellebores, until Mr. Salisbury very properly separated it from a family of plants, with which it wholly disagrees in habit, and constituted a new genus by the name of *Coptis*. This genus is characterized by the following marks. *Calyx none; petals five or six, caducous; nectaries five or six, cucullate; capsules from five to eight, pedicelled, beaked, many seeded.* The species *trifolia* has *ternate leaves,* and *a one flowered scape.*

* "Non sine admiratione vidi non solum multas cum rarissimis nostris plantis Lapponicis communes, sed etiam alias, partim ignotas omnino, partim minime tritas; et denique quasdam etiam cum Canadensibus easdem, argumento Canadam a Camscatca non longe distare, uti sequentes antea in sola America boreali visæ, nunc etiam in extrema ora Siberiæ." *Amœnitates Academicæ*, ii. 310.

In botanical arrangements, the Coptis will follow the Hellebores, from which it was taken, remaining in the class and order *Polyandria*, *Polygynia*, with the Multisiliquæ of Linnæus and the Ranunculaceæ of Jussieu.

The roots of this plant, from which the name of *goldthread* is taken, are perennial and creeping. On removing the moss and decayed leaves from the surface of the ground, they discover themselves of a bright yellow colour, running in every direction. The bases of the new stems are invested with a number of yellowish, ovate, acuminate stipules. Leaves ternate, on long slender petioles; leafets roundish, acute at base, lobed and crenate, the crenatures acuminate; smooth, firm, veiny. Scape slender, round, bearing one small, starry white flower, and a minute, ovate, acute bracte at some distance below. Calyx none. Petals five, six or seven, oblong, concave, white. Nectaries five or six, inversely conical, hollow, yellow at the mouth. Stamens numerous, white, with capillary filaments and roundish anthers. Germs from five to seven, stipitate, oblong, compressed; styles recurved. Capsules pedicelled, umbelled, oblong, compressed, beaked, with numerous black oval seeds attached to the inner side.

The root of this plant is a pure intense bitter,

scarcely modified by any other taste. In distillation it communicates no decided sensible quality to water. The constituent with which it most abounds is a bitter extractive matter, soluble both in water and alcohol. It seems destitute of resinous or gummy portions, since the residuum from an evaporated solution in alcohol is readily dissolved in water, and vice versa. It is devoid of astringency when chewed in the mouth, and it gives no indication of the presence of tannin or gallic acid when tested with animal gelatin, or with sulphate of iron. The abundance of the bitter principle is evinced by the acetate of lead and nitrate of silver, both of which throw down a copious precipitate. The sulphuric, nitric, and muriatic acids occasion no change, and the muriate of tin gives only a slight precipitate, after some time standing.

Of this article larger quantities are sold in the druggists' shops in Boston, than of almost any indigenous production. The demand for it arises from its supposed efficacy as a local application in aphthous, and other ulcerations of the mouth. Its reputation however in these cases is wholly unmerited, since it possesses no astringent or stimulating quality, by which it can act on the ulcerated spots, and where benefit has attended its use, it is doubtless to be ascribed to other articles possessing the above properties, with which it is usually combined.

As a pure tonic bitter, capable of strengthening the viscera and promoting digestion, it is entitled to rank with most articles of that kind now in use. Its character resembles that of Gentian, Quassia, and Columbo, being a simple bitter without aroma or astringency. The tincture, made by digesting half an ounce of the bruised root in eight ounces of diluted alcohol, forms a preparation of a fine yellow colour, possessing the whole bitterness of the plant. I have given it in various instances to dyspeptics and convalescents, who have generally expressed satisfaction from its effects; at least, as frequently as from other medicines of its class. A teaspoonful may be taken three times a day. In substance, it rests well on the stomach in doses of ten or twenty grains. It is however difficult to reduce to powder on account of the tenacity of its fibres.

BOTANICAL REFERENCES.

Coptis trifolia SALISBURY, *Lin. Trans.* viii. 305.—PURSH, ii. 390.—Helleborus trifolius, *sp. pl.*—WILLD. ii. 1338. KALM, *Travels*, iii. 379.—LEPECH. *iter* i. 190.—PALLAS, *Iter*. iii. 34.—OEDER, *F. Dan. t.* 566.—MICHAUX, *Fl.* i. 325.—*Amœn. Acad.* ii. 356, *t.* 4. *f.* 18.

MEDICAL REFERENCES.

Helleborus trifolius, BART. *Coll.* Nigella.—CUTLER, *Amer. Acad.* i. 457.—THACHER, *Disp.* 283.

PLATE V.

Fig. 1. *Coptis trifolia with the root, leaves, flowers and last year's fruit.*

Fig. 2. *Nectaries, stamens, and pistils magnified.*

Fig. 3. *Section of a capsule shewing the seeds.*

ARBUTUS UVA URSI.

Bear berry.

PLATE VI.

Few shrubs are more extensively diffused throughout the northern hemisphere, both in the old and new continents, than this trailing evergreen. We are told that it abounds in the northern parts of Europe, in Sweden, Lapland, and Iceland, and extends southerly to the shores of the Mediterranean. In Siberia it is also found, and is represented as abundant on the banks of the Wolga. In North America it grows from Hudson's bay as far south, at least, as the central parts of the United States. It occupies the most barren places, such as gravelly hills and dry, sandy woods, and covers the ground with beds of considerable extent.

Fig. 1.

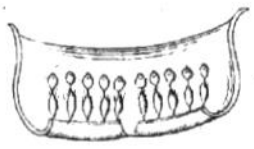

Fig. II.

Fig. III.

Fig. IV.

Fig. V.

Arbutus Uva ursi

The family of plants bearing the name of Arbutus have for their distinctive marks *a five-parted calyx, an ovate corolla, pellucid at base;* and *a superior, five-celled berry.* They are closely connected to the Vaccinia or whortleberries, from which they differ principally in the situation of the berry, which in the Arbutus grows above the calyx, and in the Vaccinium below it.—Both these genera, at least the American species, properly belong to the class *Decandria* and order *Monogynia.* The Linnæan natural order is *Bicornes.* Jussieu has them among his *Ericæ.*

The species Uva ursi, Bear's grape or Bearberry is known from the rest by its procumbent stem and entire leaves.—It trails upon the ground, putting out roots from the principal stems, and tending upward with the young shoots only. The cuticle is deciduous, and peels off from the old stems. Leaves scattered, obovate, acute at base, attached by short petioles, coriaceous, evergreen, glabrous, shining above, paler beneath, entire, the margin rounded, but scarcely reflexed, and in the young ones pubescent. Flowers in a short cluster on the ends of the branches. Peduncles reflexed, furnished at base with a short acute bracte underneath, and two minute ones at the sides. Calyx of five roundish segments, of a reddish colour and persistent. Corolla ovate or urceolate, white with a reddish tinge, transparent at base, contracted at the mouth, hairy inside, with five short, reflexed segments. Stamens inserted at the base of the corolla with hairy filaments, and anthers with two horns and two pores in each. Germ round, style straight, longer than the stamens, stigma simple. Nectary a black indented ring, situated below the germ, and remaining till the fruit is ripe. Berries globular, depressed, of a deep red, approaching scarlet, containing an insipid, mealy pulp, and about five seeds, which in the American plant cohere strongly together, so as to appear like the nucleus of a drupe.

The leaves and stems of the Uva ursi are used in Sweden and Russia for the purpose of tanning leather. According to Linnæus, large quantities are annually collected for this use.

When chewed in the mouth, the leaves have an astringent taste, combined with some degree of bitterness. The result of such chemical trials as I have made with them, shews that they abound in tannin, which is probably their chief active constituent. A solution of gelatin occasions a copious precipitate; sulphate of iron an equally copious one of a black colour. Nitrate of mercury and lime water gave large precipitates from the decoction, the first of a light green, the last of a brownish colour. Of the existence of gallic acid, at least as it exists in galls, I have found no sufficient proof. The decoction does not redden vegetable blues, and the black precipitate with the sulphate of iron soon subsides, leaving the fluid nearly colourless. The quantity of resin, mucous matter and extractive, provided they exist in this plant, must be minute ; since the decoction was not rendered turbid by the addition of alcohol or ether, nor the tincture by the addition of water, although after standing twenty four hours, some slight flocculi appeared. Muriate of tin produced no precipitation from the decoction, though it gave one from the tincture. Acetite of lead and nitrate of silver gave large precipitates. Water distilled from this plant, suffered no change with sulphate of iron, or muriate of tin.

Professor Murray of Gottingen, finding a greater amount of soluble matter taken up by water than by alcohol, considers the former as the best menstruum for this article. A similar inference from the American plant was made by Dr. John S. Mitchell in an inaugural dissertation, published at Philadelphia in 1803. For medical uses, Murray prefers the decoction to the infusion.

The Uva ursi was probably known to the ancients, as it grows in all the southern parts of Europe. Clusius thinks it was the αρκτου σταφυλη of Galen, celebrated by him as a remedy in hemoptysis, and described as follows. "Uva ursi in Ponto nascitur, planta humilis et fruticosa, folio Memæcyli, fructum ferens rubrum, rotundum, gustu austerum." But it is well known that the brief and imperfect descriptions of the ancients were productive of little else than uncertainty in Botany.

In modern times the Uva ursi was brought into notice about the middle of the eighteenth century by De Haen, as an efficient remedy in nephritic and even in calculous cases. It had been previously in use for these complaints in Spain, at Naples and Montpellier, and as a general astringent, at a still earlier period. Its reputation was still further augmented by subsequent dissertations, published upon its properties, and different sets of experiments were instituted to ascertain if it were not actually capable of dissolving the stone of the bladder. The results most in favour of its solvent power were those of Girardi, who diminished the weight and consistency of urinary calculi, by digesting them in a preparation of this plant. It appears however that the preparation, which he employed, was an acid liquor, obtained by a destructive distillation of the leaves, and probably not superior to other weak acids in its solvent powers. On the other hand, Professor Murray found what might reasonably be expected, that these calculi were not materially affected by long digestion in a decoction of this plant at various temperatures.

The attention of many medical writers has been called to the properties of this medicine, and their reports as to its success are extremely various. Among its greatest friends, are De Haen, Professor Murray, and Dr. Ferriar; while of those whose opinion is more unfavourable, are Sauvages, Haller, Donald, Munro and Fothergill. Dr. Cullen adopts the opinion of De Heucher, that the symptoms of calculus generally are susceptible of relief from astringents, and believes that on this principle the Uva ursi is capable of mitigating complaints arising from that source.*

In this country the Uva ursi has acquired the good opinion of practitioners of medicine in re-

* In the preface to the third volume of Medical Observations and Inquiries, published at London, it is stated in very general terms, that the Uva ursi had been prescribed *unsuccessfully* by many of the members of the Society of Physicians in London. Dr. Woodville, in his Medical Botany, has unfortunately misquoted this passage, by reading "successfully" instead of "unsuccessfully."

peated instances. Professor Wistar of Philadelphia, as cited by Dr. Mitchell, has in several cases found *symptoms* like those of urinary calculus completely removed by this medicine. But these could not probably have been cases of real calculus. The late Professor Barton found the plant of much service in his own case of nephritic paroxysms, alternating with gout in the feet.

From the various testimonies which have been given respecting the properties of this article, we are not warranted in believing it to possess any real lithontriptic power. At the same time it undoubtedly proves a palliative for calculous symptoms in many cases.

I have repeatedly watched its effects in paroxysms of nephritis, brought on by gravelly concretions, and am on the whole inclined to believe in its tendency to allay sensibility in these cases, and to hasten the relief of the symptoms. It ought generally to be preceded by evacuations, and may be advantageously accompanied with opium.—In cases of dysury arising from a variety of causes, I have given the decoction of this plant with very satisfactory success in repeated instances.

The other diseases in which this plant has been recommended are, catarrhus vesicæ, leucorrhæa and gonorrhæa. All these complaints it has doubtless cured, but is at the same time inferior to other medicines in use for the same purposes.

Some years ago the Uva ursi was recommended as a remedy in pulmonary consumption by Dr. Bourne of Oxford in England, and by other writers in the periodical works. It was stated to have a very sensible effect in diminishing hectic fever, and abating the frequency of the pulse dependent on it. We do not find however that subsequent experience has justified the expectations formed of it in this disease.

In Dr. Mitchell's experiments on the pulse with this medicine, it appears that the pulsations were sometimes, not always, slightly increased after taking it, but that in every case they soon sunk below the natural standard, and remained so for some time.

Of the powder of the leaves of Uva ursi, from one to two scruples may be given to most patients. Dr. Ferriar's dose in nephritis was from five to ten grains, but a larger quantity is more effectual, and is readily borne by the stomach. The decoction may be made from half an ounce of the leaves boiled for ten minutes in a pint of water. From a wine glass to a gill of this may be taken every hour.

10

BOTANICAL REFERENCES.

Arbutus Uva ursi, LINNÆUS, *Fl. Lapponica*, 162, *t.* vi. *f.* 3.—OEDER, *Fl. Dan. t.* 33.—WOODVILLE, i. *t.* 70.—SMITH, *Fl. Brit.* 443.—*Engl. Bot. t.* 714.—MICHAUX, *Fl.* i. 249.—PURSH, i. 282.—Uva ursi, J. BAUHIN, i. 523.—CLUSIUS, *Hispan.* 79.—LOBEL, *Icon.* i. 366.—PARKINSON, *theatr.* 1457.—Vitis Idæa, RAIUS, *Hist.* 1489.

MEDICAL REFERENCES.

MURRAY, *Apparatus Med.* ii. 64.—GIRARDI, *de Uva ursina*, &c.—DE HAEN, *Ratio medendi*, ii. 160, &c.—SAUVAGES, *Nosol.* iii. 2, 200.—D. MUNRO, *Mat. Med.* iii. 288.—FOTHERGILL, *Med. Obs.* 144.—ALEXANDER, *Exp. essays*, 151.—FERRIAR i. 109.—HEBERDEN 79, 360.—DAVIE, *Med. and Phys. Journal*, xv. 347.—BOURNE, *in ditto*, xiv. 463.—SCHŒPF, 67.—MITCHELL, *Inaugural Thesis.*

PLATE VI.

Fig. 1. *Arbutus Uva ursi, the American variety.*

Fig. 2. *The magnified corolla opened, shewing the insertion of the stamens.*

Fig. 3. *Calyx, nectary, germ, and style magnified.*

Fig. 4. *Calyx and nectary.*

Fig. 5. *Berry.*

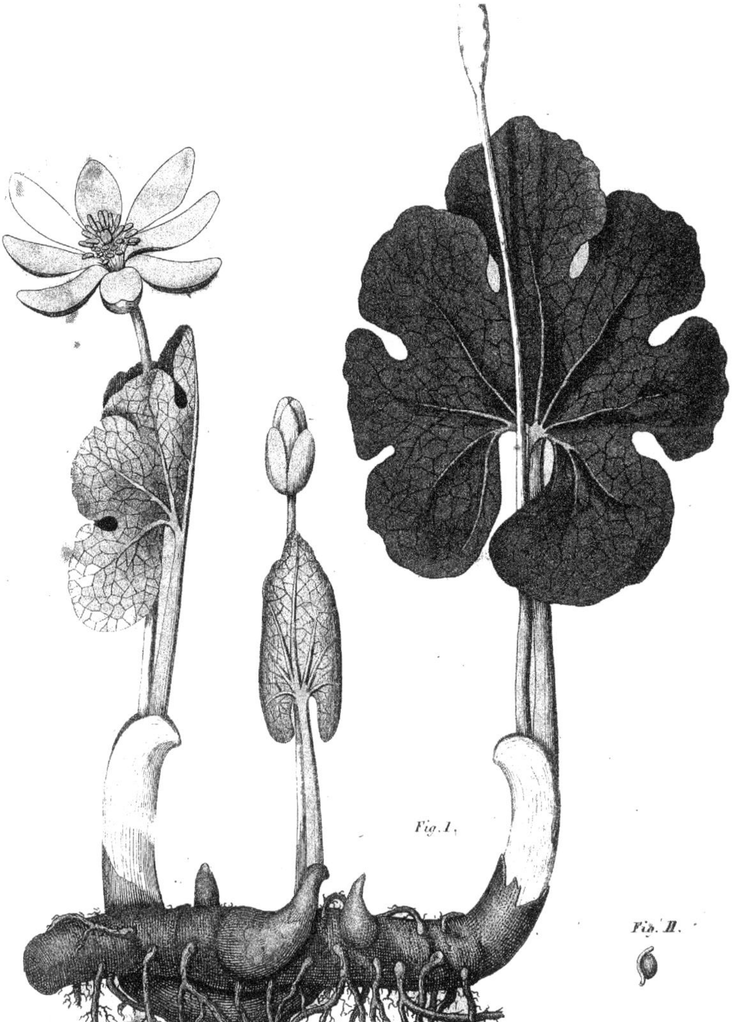

Sanguinaria Canadensis

SANGUINARIA CANADENSIS.

Blood root.

PLATE VII.

Among the earliest visiters of spring the botanist will find in almost any part of the United States the *Sanguinaria Canadensis.* Its fine white flowers proceeding from the bosom of a young, convoluted leaf, become visible in the woods, in Carolina, in the month of March, and in New England, toward the end of April. Its most common name is *Blood root.* It has also the appellation of *Puccoon, Turmeric, Red root, &c.* It is the only species we at present possess of the genus Sanguinaria, distinguished by a *two leaved calyx eight petals, and an oblong capsule, with one cell and many seeds.*—Class *Polyandria*, order *Monogynia.* Natural order *Rhœadeæ, L. Papaveraceæ,* Juss.

The flower and leaf proceed from the end of a horizontal, fleshy, abrupt root, fed by numerous radicles. This root makes offsets from its sides, which separate as the old root decays, acquiring by this separation the abrupt or premorse form.

Externally the colour of the root is a brownish red. Internally it is pale, and when divided emits a bright orange coloured juice from numerous points of its surface. The bud or hybernaculum, which terminates the root, is composed of successive scales or sheaths, the last of which acquires a considerable size, as the plant springs up. By dissecting this hybernaculum in the summer or autumn, we may discover the embryo leaf and flower of the succeeding spring, and with a common magnifier, even the stamens may be counted.

The Sanguinaria is smooth throughout. The leaves grow on long channelled petioles. When spread out, they are reniform or heart shaped, with large roundish lobes separated by obtuse sinuses. The under side is strongly reticulated with veins; it is paler than the upper, and at length becomes glaucous. The scape is round, rises in front of the petiole, and is infolded by the young leaf. The calyx consists of two concave, ovate, obtuse leaves, which are perfect in the bud, but fall off when the corolla expands. Petals eight, spreading, concave, obtuse, the alternate or external ones longer, so that the flower has a square appearance. This is its natural character, although cultivation sometimes increases the number of petals. Stamens numerous, with oblong yellow anthers. Germ oblong, compressed, style none, stigma thick, somewhat two lobed. Capsule oblong, acute at both extremities, two valved. Seeds numerous, roundish, compressed, dark shining red, half surrounded with a peculiar white vermiform appendage, which projects at the lower end.

After the flower has fallen, the leaves continue to grow, and by midsummer have acquired so large a size as to appear like a different plant.

The root of this vegetable is the only part which I have submitted to chemical examination. The experiments made on this substance, gave evidence of the following constituent principles.

1. A peculiar resin. Alcohol comes off from the root strongly impregnated with its colour and taste. This solution is rendered turbid by the addition of water. When evaporated to dryness, it leaves a residuum partially, but not wholly soluble in water. When successive quantities of water have been agitated with the powdered root until the infusion comes off colourless, alcohol acquires

a colour from the remainder. Æther receives from the root a yellowish colour, and when evaporated, leaves the resin nearly pure. In this state it is moderately adhesive, of a deep orange colour, bitter and acrid, diffusible, but not soluble in water. The resin may also be precipitated in small quantities from alcohol by water.

2. A bitter principle. Both water and alcohol acquire a strong bitter taste when digested on the root. From both these solutions a copious precipitate is thrown down by the nitrate of silver and the acetite of lead. Muriate of tin gradually renders the solution turbid, but without a precipitate. Oxymuriatic acid renders the alcoholic solution turbid, but produces no change in the watery solution for some time. At length a precipitate forms and slowly subsides; but produces no change in the watery solution. No precipitate was formed from the cold aqueous infusion in an hour by the sulphuric or nitric acids, by lime water, nitrate of mercury, muriate of barytes, oxalate of ammonia, sulphate of iron, gelatine or hydro-sulphuret of potash. After standing twenty four hours, a very slight precipitate was discovered from the lime water and nitrate of mercury only.

3. An acrid principle. The acrimony resides in part in the resin, but is also communicated to

water. It is diminished by heat, yet it does not come over with water in distillation.

4. Fæcula. The infusion of the root in cold water is limpid. The hot infusion is viscid and glutinous and stiffens linen. From this solution the fæcula is precipitated in a white powder by alcohol. Nitric acid dissolves this precipitate, which may be again thrown down by alcohol.

5. A fibrous or woody portion.

The beautiful colour of the root seems to reside more in the resin than in any other principle, since the alcoholic solution has always more than twice as much colour as the aqueous. Papers dipt in these solutions receive a bright salmon colour from the tincture, but a very faint one from the aqueous infusion. This circumstance furnishes an impediment to the use of this article in dyeing.

The medical properties of the Sanguinaria are those of an acrid narcotic. When taken in a large dose it irritates the fauces, leaving an impression in the throat for considerable time after it is swallowed. It occasions heartburn, nausea, faintness, and frequently vertigo and diminished vision. At length it vomits, but in this operation it is less certain than other emetics in common use. The above effects are produced by a dose of from eight to twenty grains of the fresh powdered root.

When given in smaller doses, such as produce nausea without vomiting, and repeated at frequent intervals, it lessens the frequency of the pulse in a manner somewhat analogous to the operation of Digitalis. This however is a secondary effect, since in its primary operation it seems to accelerate the circulation. Exhibited in this manner, it has been found useful in several diseases.

In still smaller doses, or such as do not excite nausea, it has acquired some reputation as a tonic stimulant.

Professor Smith of Hanover, New Hampshire, in a paper on this plant, published in the London Medical Transactions, vol. i. states that he found the powder to operate violently as an emetic, producing great prostration of strength, during its operation, which continued for some time. He had not known it to act as a cathartic. Snuffed up the nostrils, it proved sternutatory, and left a sensation of heat for some time. Applied to fungous flesh it proved escharotic, and several polypi of the soft kind were cured by it in his hands. He found it of great use in the incipient stages of pulmonary consumption, given in as large doses as the stomach would bear, and repeated. In cases of great irritation it was combined with opium. Some other complaints were benefitted by it, such as acute rheumatism and jaundice.

Professor Ives of New Haven* considers the Blood root as a remedy of importance in many diseases, particularly of the lungs and liver. He observes, that in typhoid pneumonia, "in plethoric constitutions, when respiration is very difficult, the cheeks and hands become livid, the pulse full soft, vibrating and easily compressed,—the Blood root has done more to obviate the symptoms and remove the disease," than any remedy which he has used. In such cases, he observes, "the dose must be large in proportion to the violence of the disease, and often repeated, until it excites vomiting, or relieves the symptoms." He infuses from a scruple to half a drachm of the powdered root in half a gill of hot water, and gives one or two teaspoonfuls every half hour, in urgent cases, until the effect is produced. This treatment has often removed the symptoms in a few hours.

Dr. Ives thinks highly of its use in influenza, in phthisis, and particularly in hooping cough. He also states, that given in large doses, sufficient to produce full vomiting, it often removes the Croup, if administered in the first stages. It has been given, he remarks, "for many years in the

* Letter dated November 5, 1816.

country, some physicians relying wholly on this remedy for the cure of croup."

Dr. Macbride, of Charleston, S. C. who has contributed many judicious remarks on the medicinal properties of plants, to Mr. Elliott's excellent Botany of the Southern States; informs me,* that he has found the Blood root useful in Hydrothorax, given in doses of sixty drops, *ter de die*, and increased until nausea followed each dose. In a week or two the good effect was evident, the pulse being rendered slow and regular, and the respiration much improved. In the same letter he observes, " In torpor of the liver, attended with colic and yellowness of the skin, a disease common in this climate, I use the Puccoon with evident advantage. We use it also in jaundice, but in this disease I do not trust exclusively to it. I prefer the pill or powder (dose from two to five grains) and vinous infusion, to the spirituous tincture."

The tincture of Sanguinaria may be made by digesting an ounce of the powdered root in eight ounces of diluted alcohol. This preparation possesses all the bitterness, but less of the nauseating quality, than the infusion. In the dose of a small teaspoonful, it is used by many practitioners

* Letter dated December, 1816.

as a stimulating tonic, capable of increasing the appetite and promoting digestion.

BOTANICAL REFERENCES.

Sanguinaria Canadensis, Lin. *sp. pl.*—Curtis, *Botan. Mag.* *t.* 162.—Aiton, *Hort. Kew.* ii. 222.—Walter, *Carol.* 153.—Michaux, *Flora* 1, 309.—Pursh, ii. 366.—Sanguinaria minor, Dillenius, *Elth. f.* 326 *and* S. major, *f.* 325 *in t.* 252.—Chelidonium maximum acaulon Canadense Raius, *Hist.* 1887.—Ranunculus Virg. albus. Parkinson, *Th.* 326.—Chelidonium majus Canad. acaulon Cornutus, *Canad.* 212.

MEDICAL REFERENCES.

Schœpf, 85.—Smith, *Trans. Lond. Med. Society*, i. 179.—Bart. *Coll.* 28.—Cutler, *Mem. Amer. Acad.* i. 455.—Thacher, *Disp.* 331.

GERANIUM MACULATUM.

Common Cranesbill.

PLATE VIII.

In common language the term Geranium includes all that extensive tribe of plants comprised by the old genus of that name, and principally characterised by their beaked fruit and five seeds which are scattered by means of awns. L'Heritier has divided this family into three distinct genera, under different orders in the artificial class *Monadelphia.* These are *Erodium,* having five stamens, five nectariferous scales and glands, and the awns of the fruit twisted and bearded. *Pelargonium,* which includes most of the Cape species so commonly cultivated among us, having about seven stamens, an irregular corolla, and a nectareous tube running down the peduncle. Lastly, *Geranium* having ten stamens, a regular corolla, five nec-

Fig. I.

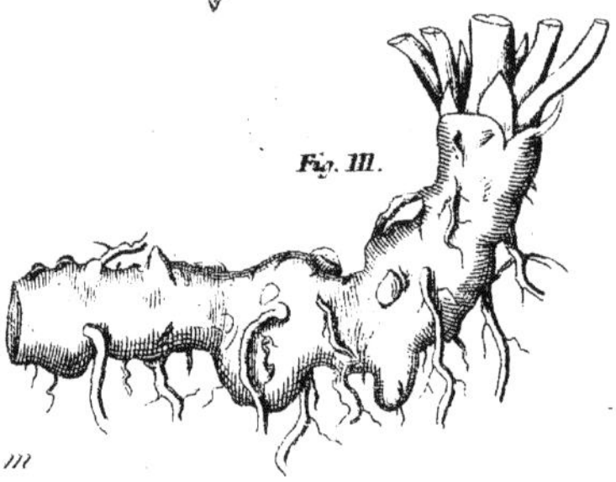

Fig. III.

Fig IV

ranium maculatum

tariferous glands at the base of the longer filaments, the awns of the fruit neither bearded nor twisted. To this division belongs the plant under consideration, which has the following specific character. *Erect, hairy backward; stem forked; leaves opposite, three or five parted, cut; peduncles mostly two flowered; petals, obovate, entire.*

Jussieu has formed a natural order by the name of *Gerania*, which nearly corresponds to the *Gruinales* of Linnæus.

Although we have few species of Geranium in the United States, yet the present species, by its extensive diffusion, is a sufficient representative of the race. It is very common in low grounds, about Boston and Philadelphia, in the Carolinas, and in the western country upon the banks of the Ohio and Illinois.

The root of Geranium maculatum is perennial, horizontal, thick, rough and knobby. In most plants it sends up a stem and several root leaves. The leaves are spreading, hairy, divided in a palmate manner into three, five, or seven lobes, which are variously cut and toothed at their extremities; those of the root are on long petioles, those at the middle of the stem opposite and petioled, those at the top opposite and nearly sessile. The stem is erect, round, hispid with reversed hairs, dichotomous, with a flower stalk in the fork. Stipules and bractes linear, dilated at base. Peduncles round, hairy, swelling at base, generally two flowered. Calyx of five oblong, ribbed, mucronated leaves, with the parts, which are outermost in the bud, hairy. Petals five, obovate, not emarginate, of a light purple colour, which is deeper when the plant grows in the shade, marked with green at the base. Stamens ten, erect or curving outward, the alternate ones a little longer, with nectariferous glands at the base; filaments dilated and united together at base; anthers oblong, deciduous, so that the number frequently appears less than ten. Germ ovate; style straight, as long as the stamens; stigmas five, at first erect, afterwards recurved. Capsule five seeded, surmounted by a long straight beak, from the sides of which when ripe are separated five thin, flat awns, which curl up, having cast off the seed contained in the cell at the base of each.

The root of the Geranium, which is the part to be used in medicine, is internally of a green colour, and when dry is exceedingly brittle and easily reduced to powder. It is one of the most powerful astringents we possess, and from its decided properties, as well as the ease of procuring it, it may well supersede in medicine many foreign articles of its class which are consumed among us. The experiments, which I have made upon this root, have been principally directed to the examination of its astringent qualities.

A drachm of the powdered root was steeped in two ounces of cold water and the infusion filtrated. Successive portions of water were added until the liquid came off colourless and tasteless. The collected infusion had a pale greenish colour, and a styptic, austere taste. It did not redden vegetable blues.

To half this infusion was added a drachm of gelatin in solution. The liquor instantly became of a milky whiteness, and a copious white precipitate was thrown down. This precipitate was dried and assumed a semi-transparent, horny appearance. Its weight was eleven grains.

A drachm of kino treated in the same manner was rendered turbid, but gave a very scanty precipitate with the gelatin.

To portions of the same infusions was added a solution of the muriate of tin. In both of them a greenish precipitate was formed, but that of the Geranium was much the most immediate and abundant.

The sulphate of iron struck a dark purple colour with the infusion of Geranium. The compound remained principally suspended at the end of twenty four hours, and when used in writing had the appearance of common ink, but in a few days changed to a dull brown colour. A portion of the fresh infusion was distilled, but the liquid which came over was not altered in colour by the sulphate of iron.

The above experiments indicate the presence of tannin and gallic acid, the former in large quantities, in the root of the Geranium. The proportion of tannin seems considerably to exceed that in the kino of the shops. The gallic acid is indicated by the dark precipitate remaining in solution. This is Berthollet's criterion. It differs however from the acid of oak galls in not reddening vegetable blues, and not passing over in distillation.

Alcohol and proof spirit readily dissolve the active constituents of the Geranium. The tincture has a great sensible astringency, and is a convenient mode of keeping the article for use.

The Geranium has been repeatedly employed in medicine by various practitioners in this country. I have found it useful in a number of cases, where astringents were capable of rendering service. It is particularly suited to the treatment of such discharges as continue from debility after the removal of their exciting causes. The tincture forms an excellent local application in sore throats and ulcerations of the mouth.

Its internal use has been recommended in dysentery and cholera infantum, but astringents are not always admissible in these complaints, at least in their early stages, during the existence of much active inflammation, or during the presence of any substance requiring to be removed.

The Geranium may be used in powder in extract, or in tincture. Its doses are similar to those of kino and catechu, a drachm or two of the tincture, twenty or thirty grains of the powder, and a quantity somewhat less of the extract.

BOTANICAL REFERENCES.

Geranium maculatum, *Sp. pl.* WILLDENOW, iii. 705.—GRONOVIUS, *Virg.* 101.—WALTER, *Carol.* 175.—MICHAUX, ii. 38.—PURSH, ii. 448.—G. caule erecto, herbaceo, foliis oppositis, quinque partitis, incisis &c. CAVANILLES, *diss. t.* 86, *f.* 2.—G. batrachioides, Americanum, maculatum, floribus obsolete cœruleis. DILL. *Elth.* 158. *t.* 131, *f.* 159.

MEDICAL REFERENCES.

SCHŒPF, 107.—BART. *Coll.* 7.—CUTLER, *Mem. Amer. Acad.* i. 469.—THACHER, *Disp.* 224.

PLATE VIII.

Fig. 1. *Geranium maculatum.*
Fig. 2. *The fruit.*
Fig. 3. *The root.*

TRIOSTEUM PERFOLIATUM.

Fever root.

PLATE IX.

THIS is rather a solitary plant, and though met with in most parts of the United States, it rarely, I believe, occurs in large quantities. About Boston it is found in several places at the borders of woods in rich, shady situations. Its common names are *Fever root* and *Wild ipecac.** Pursh observes, that it is rare, and generally occurs in limestone soils. With us it flowers in June and ripens its fruit in September.

The genus *Triosteum* is found in the class

* The quaint appellation of Dr. *Tinker's weed*, which has been bestowed on this plant, is thus gravely commented on by Poiret. "Ses racines et celles de l'espece précédente passent pour émétiques ; le docteur Tinkar est le premier qui les a mises en usage, et qui a fait donner à cette plante par plusieurs habitans de l'Amerique septentrionale le nom d' *herbe sauvage du docteur Tinkar.*"

Fig. 1.

Fig. II.

Fig. III

Fig IV

Fig V.

Fig

Triosteum perfoliatum

Pentandria and order *Monogynia.* Its natural affinities place it among the *Aggregatæ* of Linnæus and the *Caprifolia* of Jussieu. It is characterized by a *monopetalous, five-lobed, unequal corolla; a calyx as long as the corolla; and a berry with three cells and three seeds.* The species *perfoliatum* differs from the rest in having its *leaves connate, and its flowers sessile and whorled.*

The root of this plant is perennial and subdivided into numerous horizontal branches. The stem is erect, hairy, fistulous, round, from one to four feet high. The leaves are opposite, the pairs crossing each other, connate, ovate, acuminate, entire, rather flat, abruptly contracted at base into a sort of neck, resembling a winged petiole. This portion varies in width, as Michaux has expressed it, "foliis latius, angustiusve con natis." In general it is narrow when the plant is in flower, as represented in the figure; and wider when it is in fruit. The flowers are axillary, sessile, five or six in a whorl, the upper ones generally in a single pair. Each axil is furnished with two or three linear bractes. The calyx consists of five segments which are spreading, oblong-linear coloured, unequal, persistent. Corolla tubular, curving, of a dull brownish purple, covered with minute hairs, its base gibbous, its border open and divided into five rounded, unequal lobes. Stamens inserted in the tube of the corolla, hairy, with oblong anthers. Germ inferior, roundish; style longer than the corolla; stigma peltate. The fruit is an oval berry of a deep orange yellow,* hairy, somewhat three sided, crowned with the calyx, containing three cells and three hard, bony, furrowed seeds, from which the name of the genus is taken.

This plant was made the subject of an interesting communication to the Linnæan society of New England, by Dr. John Randall. The experiments made by him on its medical uses and pharmaceutical preparations were numerous, and serve to throw much light on its properties. In trying the solvent powers of water and alcohol, he found that water afforded a much greater quantity of extract than alcohol, and that the spirituous extract was perfectly soluble in water, whence he infers that no resin in a pure state exists in the plant. He discovered no volatile oil by distillation, nor any other principle of activity in water distilled from the plant. He concludes also, that

* Pursh observes that the flowers and berries are *purple*. In all the specimens I have examined, which have not been few in number, the fruit was of a bright orange colour. If Pursh has seen a plant with purple berries, it is probably a different species from the true plant of Linnæus and Dillenius, which had "*fructus lutescentes*."

no free acid exists in this vegetable. Of the different parts submitted to examination, the leaves yielded the greatest quantity of soluble matter, but the root afforded that of the greatest activity. By decoction and evaporation with water an ounce of the dried stalks afforded one drachm of extract; an ounce of the dry roots, two drachms and two scruples, and the same quantity of leaves half an ounce. From a similar treatment of equal portions with alcohol, rather more than half the above quantities of extract were obtained.

The sensible qualities of the root were found essentially different from those of the herb. Both of them possess a large share of bitterness, but the root has also a nauseous taste and smell, somewhat approaching to those of ipecacuanha. The medical properties of the Triosteum are those of an emetic and cathartic. In the above dissertation, about thirty cases are detailed, in which different preparations and quantities of the article were given to various persons with a view to their medicinal effects. The general inference to be made from them is, that the bark of the root acts with tolerable certainty as an evacuant upon the alimentary canal, both by emesis and catharsis. When given alone, either in powder or decoction, the instances of its failure were not many, and when combined with calomel, its operation was attended with a certainty, hardly inferior to that of jalap. The aqueous and spirituous extract of the root were likewise efficacious, and nearly in an equal degree. Preparations made from different parts of the herb possessed much less activity, the decoction of the leaves operating only as a diaphoretic, and that of the stalk producing no effect.

The late Professor Barton of Philadelphia, in his Collections toward a Materia Medica of the United States, speaks of this plant as a mild and good cathartic, sometimes operating as a diuretic and in large doses as an emetic.

My own experience with this plant has not been extensive, yet sufficient to satisfy me of its medicinal power. Where I have administered it, it has generally proved cathartic, a larger dose however being requisite for this purpose, than of jalap or aloes. It has sometimes failed to produce any effect, and I am inclined to believe that its efficacy is much impaired by age. Those who may incline to employ it, will do well to renew their stock annually, and to keep the powder in close stopped phials.

A dose of the bark of the root in powder is twenty or twenty five grains, and of the extract, a somewhat smaller quantity.

BOTANICAL REFERENCES.

Triosteum perfoliatum, Lin. *sp. pl.* Aiton, *Hort. Kew*, i. 234.—Pursh. i. 162.—Triosteum majus, Michaux, *Fl.* i. 107.—T. floribus verticillatis, sessilibus, Gronov. 31.—Triosteospermum latiore folio, flore rutilo, Dillenius, *Elth. t.* 293. *f.* 378.

MEDICAL REFERENCES.

Schœpf, 23.—Bart. *Coll.* 29.

PLATE IX.

Fig. 1. *Triosteum perfoliatum.*

Fig. 2. *A flower separated.*

Fig. 3. *The corolla opened, shewing the stamens and style.*

Fig. 4. *The calyx.*

Fig. 5. *The fruit, crowned with the calyx.*

Fig. 6. *The same dissected to shew the three seeds.*

Fig. 7. *A seed.*

RHUS VERNIX.

Poison Sumach or Dogwood.

PLATE X.

THE fine, smooth foliage of the Rhus vernix render it one of the most elegant of our native shrubs, while its well known poisonous qualities make it an object of aversion, and deter most persons from a near inspection of its structure and characteristics. From Canada to Carolina it is a common tenant of swamps and meadows, usually attaining the height of ten or fifteen feet, but sometimes rising into a tree of twice that altitude. The names of *Poison tree*, *Poison wood*, *Poison ash*, &c. are applied to it in different parts of the United States. In Massachusetts it is universally known by the name of *Dogwood.* This appellation, being applied throughout the country to *Cornus florida*, serves to shew the fallacy of de-

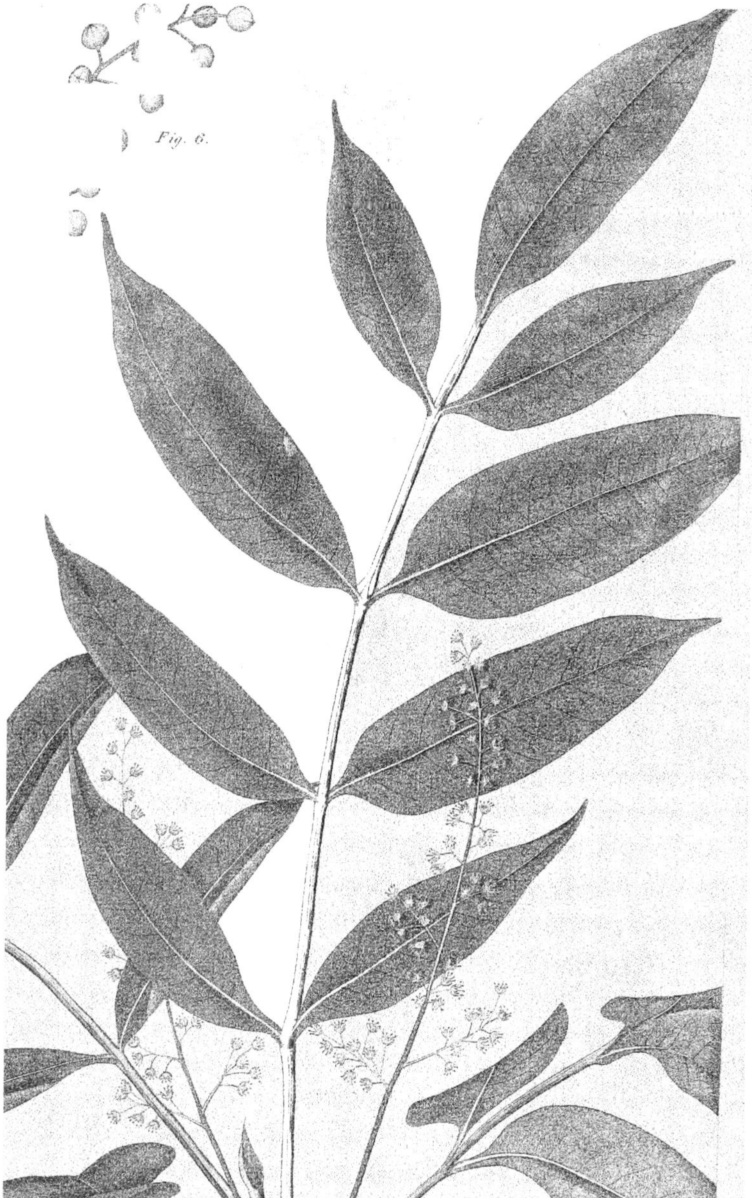

Rhus vernix

pending on vulgar or provincial names for the distinction of plants. A mistake of very injurious consequence might easily arise from the confusion of the English names of two trees so dissimilar in their qualities.

The class *Pentandria* and order *Trigynia;* the Linnæan order *Dumosæ* and the Jussieuean *Terebintaceæ* include the genus to which this shrub belongs. The generic character consists in *an inferior, five-parted calyx, a corolla of five petals, and a berry with one seed.* The Rhus vernix has its leaves *annual, pinnate, glabrous;* its *leafets oblong, entire, acuminate; its panicle lax, and its flowers diœcious.*

The trunk of the poison sumach is from one to five inches in diameter, branching at top, and covered with a pale greyish bark. The wood is light and brittle, and contains much pith. The ends of the young shoots and the petioles are usually of a fine red colour, which contributes much to the beauty of the shrub. The leaves are pinnate, the leafets oblong or oval, entire, or sometimes slightly sinuate, acuminate, smooth, paler underneath, nearly sessile, except the terminal one. The flowers, which appear in June, are very small, green, in loose axillary panicles. Where they appear not axillary, it is because the leaf under them has been detached. The barren and fertile flowers grow on different trees. The panicles of barren flowers are the largest and most branched. They are furnished with short, oblong bractes, and downy pedicels. The calyx has five ovate segments, and the corolla five oblong, sigmoid petals. The stamens are longer than the petals, and project through their interstices. The rudiment of a three-cleft style is found in the centre.—In the fertile flowers, the panicles of which are much smaller, the calyx and petals resemble the last, while the centre is occupied by an oval germ, ending in three circular stigmas. The fruit is a bunch of dry berries or rather drupes of a greenish white, sometimes marked with slight purple veins, and becoming wrinkled when old. They are roundish, a little broadest at the upper end, and compressed; containing one white, hard, furrowed seed.

A tree, supposed to be the same with the Rhus vernix, grows in Japan, and furnishes the celebrated black varnish of that country.

A controversy respecting the identity of the Japanese and American species, was carried on in the forty ninth and fiftieth volumes of the London Philosophical transactions, by Mr. Philip Miller and Mr. John Ellis. The mass of evidence seems to justify the belief, that notwithstanding the remoteness of their situation, they are one and the same species. The description of the oriental tree, given in Kæmpfer's Amœnitates exoticæ, agrees very closely with that of the American species. [*Note* G.] Like our native Rhus, the Japanese tree possesses a poisonous influence, and frequently causes a severe cutaneous affection in those who approach or gather it. It only remains to shew, that a varnish may be obtained from the American Rhus vernix, to furnish strong presumptive evidence of the identity of the two.

If an incision be made in the bark of our Rhus vernix in the spring or autumn, a quantity of thick viscid fluid immediately exudes, and sometimes with such rapidity as to drop off before it can be collected. This juice has an opaque, whitish appearance, and a strong, penetrating, disagreeable smell. On exposure to the atmosphere, its colour soon changes to a deep black. It is extremely slow in drying, and permanently retains its black colour.

In the month of October, 1814, with the assistance of Dr. Pierson, whose case is afterwards described; I collected several ounces of this juice from a thicket of trees in Brighton. Being collected in a phial, it retained its whitish colour, except at the surface, where it turned black from its contact with the air in the upper part. This juice was kept for more than two years without any change in its appearance. In cold weather it was extremely viscid, and flowed with difficulty.

Different portions of this juice were submitted to chemical examination. It was perfectly insoluble in water, although upon boiling with it, it formed a thick emulsion. Alcohol dissolved it sparingly, and the solution was rendered turbid by water. Æther combined with it more largely, forming a thick, opaque compound. Strong sulphuric acid combined with it, producing a black solid mass. Alkalies also combined with it, and a strong solution of pearl ash dissolved a portion of it, which was afterwards precipitated by sulphuric acid. It had an affinity for metallic oxyds, and powdered litharge, upon being boiled with it, rendered it nearly solid. In distillation at the heat of boiling water, nothing came over except a slight film upon the surface of the water. When the heat was raised to the boiling point of the juice, a quantity of thin, blackish, volatile oil came over, which dried up on being exposed to the air, leaving a slight coating on the surface of the vessel which contained it. The portion remaining in the retort was much inspissated, and upon cooling became nearly solid.

Being desirous to try the effect of this juice, employed as a varnish, I applied a coating of it with a brush to different surfaces of wood, glass, tinned iron, paper, and cloth. These were exposed to the air and light during the whole of the months of July and August, at the expiration of which period they had not become dry. Each of the coatings was half fluid and adhesive, and had collected much dust. Upon the cloth and paper the juice had spread extensively, giving them an oily appearance.

Concluding from this experiment that the juice could not be usefully employed in its crude state, I endeavoured to render it more *drying* by the addition of litharge. The compound, which resulted from boiling with this oxyd, became dry in a short time, but was not distinguished for any remarkable degree of lustre.

The third and last experiment proved more satisfactory. A quantity of the juice was boiled alone, until nearly all the volatile oil had escaped, and the remainder was reduced almost to the state of a resin. In this state it was applied while warm to several substances, which after cooling exhibited the most brilliant, glossy, jet black surface. The coating appeared very durable and firm, and was not affected by moisture. It was elastic and perfectly opaque, and seemed calculated to answer the purposes of both paint and varnish.

The chemical constitution of the juice of the Rhus vernix seems, from the foregoing experiments, to be most analogous to that of the balsams, consisting chiefly of a resin and an essential oil. The oil dissipates slowly at low temperatures, approaching in this and some other respects to the character of a fixed oil. The resin, when procured in contact with the atmospheric air, is black, opaque, and solid, rendered very adhesive, and at length fluid by heat.

A very distressing, cutaneous disease, it is well known, ensues in many persons from the contact, and even from the effluvium of this shrub. The poisonous influence which produces this affection is common to several other trees and plants, such as the Poison vine or Poison ivy, *(Rhus radicans,)* the Cashew nut, *(Anacardium occidentale,)* and the Manchineel, *(Hippomane mancinella.)* Even the garden Rue, and common Oleander, are said to affect some persons in a similar manner.—The Rhus vernix is the most formidable of this tribe which is found among us, and occasionally produces the most severe effects. It is however extremely various in its action, upon persons of different idosyncrasies. Some cannot come within the atmosphere of the shrub, without suffering the most violent consequences. Others are but slightly affected by handling it, and some can even rub, chew, and swallow the leaves without the smallest inconvenience.

The most formidable cases in persons subject to this poison, usually commence within twenty four hours after the exposure. The interval is sometimes longer, but more frequently shorter. The symptoms are generally ushered in by a sense of itching and a tumefaction of the hands and face. The swelling gradually extends over various parts of the body, assuming an erysipelatous appearance. The inflamed parts become more elevated, acquiring a livid redness, attended with a painful burning sensation. Small vesicles now appear upon the surface, which extend and run into each other. They contain a transparent fluid, which by degrees becomes yellow, and at length assumes a purulent appearance. A discharge takes place from these vesicles or pustules, giving rise to a yellowish incrustation, which afterwards becomes brown. In the mean time an insupportable sensation of itching and burning is felt. The inflamed parts become excessively swollen, so that not unfrequently the eyes are closed, and the countenance assumes a shapeless and cadaverous appearance, which has been compared to that in malignant small pox. The disease is usually at its height from the fourth to the sixth day, after which the skin and incrustations begin to separate from the diseased parts, and the symptoms gradually subside. It is not common for any scars or permanent traces of the disease to remain. Notwithstanding the violent character which it sometimes assumes, I never knew an authenticated case of its terminating fatally. It is however capable of occasioning the most distressing symptoms. Kalm, in his travels in North America, mentions a person who, by the simple exhalation of the Rhus vernix, was swollen to such a degree, that "he was stiff as a log of wood, and could only be turned about in sheets." Dr. Thacher mentions a case, in which the head and body were swollen to a prodigious degree, so as to occasion the loss of sight for some time; and the patient recovered at the end of several weeks with the loss of his hair and nails.

Of the cases which have fallen under my notice, the following affords a fair instance of the operation of this poison, as it ordinarily effects those who are constitutionally liable to it. On the 27th of October, 1814, Dr. A. L. Pierson, then a student of medicine, accompanied me to Brighton for the purpose of collecting the juice of the Rhus vernix, growing at that place. He had always supposed himself constitutionally exempt from liability to the poison. The day proved warm, and the effluvium from the incisions we had made in the trees was very powerful. We were engaged in the collection for upwards of an hour, during which he was less exposed than myself, being absent a part of the time. His own account of the symptoms which followed this exposure is as follows:

"I felt no unpleasant effects for six or seven hours after returning to Boston. About 8 o'clock P. M. I perceived the backs of my hands were swollen and puffy, but without pain or itching; my forehead and upper lip were soon in the same state. On the morning of the 28th the tumefaction had increased, and I discovered various other parts of my body to be infected. The backs of my hands and wrists, which were the most advanced, began to show small watery vesicles. No applications were made till the noon of this day. I then applied cloths dipped in lead water to one hand and wrist, and in a spirituous solution of the

the corrosive muriate of mercury to the other. From this and subsequent trials, I am induced to prefer the lead. The parts began to itch—the tumefaction increased—vesication began to take place on the swollen surface—small pustules formed and ran into each other, and at last some were formed as large as nutmegs. On the 29th, my eyes were nearly closed, in consequence of the swelling of my forehead, eyebrows and cheeks. The contents of the vesicles were perfectly limpid—inoculation from them to other parts had no effect—neither in this nor any subsequent stage. On the evening of the 30th, the inflammation appeared at its height. The burning sensation and itching were intolerable. I could scarcely discern any object. On the 31st, the pustules began to appear a little milky—and before night the inflammation was evidently on the decline. I this day applied an ointment, composed of Ung. Stramonii, 1 oz.—Subm. Hyd. c. Ammonia (white precipitate) 1 dr. mixed—with a very pleasant effect. It was now soothing, although before it had seemed to irritate, and produced pain when applied. November 1st, a very free desquamation began, first on my forehead, hands and wrists. And in just a fortnight I was enabled to leave my chamber, blessed with a new cuticle from the root of the hair on my forehead to my breast, from the middle of my forearm to the tips of my fingers, and on the whole inside of my thighs. The constitutional effects of this thorough vesication were but slight. During the first five days, my pulses were increased from ten to twenty strokes in the minute. The time of duration of the inflammatory symptoms in this case accords pretty well with the account of Prof. Barton, who states, I think, the height of it to be on the fifth day. It is worth observing, that the operation of the poison seemed to have a considerable effect in relieving me from dispeptic symptoms, with which I had been previously troubled, and also benefitted a chronic inflammation of my eyes. I am still subject to an eruption of watery pustules between my fingers, which dry up, and the cuticle peels off." *Letter dated July*, 1815.

Many constitutions are but slightly, or not at all, affected by the poison of the Rhus vernix. This I find to be my own case. After the same exposure, which occasioned the case just detailed, I experienced no ill consequence, except a slight vesicular eruption on the backs of the hands and about the eyes, which disappeared in a short time, without farther inconvenience. The same slight affection I have felt upon several subsequent exposures, particularly when making, from a recent specimen, the drawing which accompanies this account.

I apprehend that a majority of persons are not liable to the injurious effects of the poisonous sumacs. Among persons residing in the country, exposures must occur very frequently from the abundance of these shrubs, especially of the Rhus radicans, by roadsides and elsewhere. Very few however, in proportion to the number exposed, have personal experience of their deleterious effects. In those on the contrary, in whom a constitutional liability to the poison exists, the disease frequently returns several times during life, notwithstanding the utmost precaution in avoiding its causes. A gentleman residing in the country informed me, that he had been seven times poisoned to the most violent degree. In such constitutions a slight exposure is sufficient to excite the disease. I have known individuals badly poisoned in winter from the wood of the Rhus vernix, accidentally burnt on the fire. Others have made the same observation.

Some farther remarks on the poison of these shrubs, and on the treatment of the disease occasioned by them, will be made in a future part of this work, under the head of *Rhus radicans.*

Many interesting observations on the properties of these species of Sumach, will be found in an inaugural dissertation, by Dr. Thomas Horsfield of Bethlehem, Pa. a work of much industry and merit.

In the New York Medical Repository is an account of a swarm of bees, which, having alighted on the branches of the Rhus vernix, were the next day found dead, with their bodies black and swollen. This is a remarkable circumstance. There is certainly no instinctive aversion in these animals for the tree. In the flowering season the blossoms, which are very fragrant, are always thronged with a multitude of winged insects in quest of their honey.

The introduction of the juice into the arts will not perhaps take place among us, during the present high price of labour, and the general prejudice which exists against the shrub. In some future period, it is probable that a substance, which is found so valuable in the eastern countries, will not be neglected among us. It might safely be procured by persons not subject to the poison, and, with proper precautions, would injure no one during its preparation and use. A pound of the juice in a day might be collected by an individual. When thoroughly dry, it ceases to emit an effluvium, and nothing farther is to be apprehended from its effects.

BOTANICAL REFERENCES.

Rhus vernix LINNÆUS, *Sp. pl.*—AITON, *Hort. Kew.* i. 366.—MICHAUX, i. 183.—PURSH, i. 205.—Pennated toxicodendron ELLIS, *Phil. trans. abr.* xi. *passim.* American toxicodendron MILLER, *ibid.*—Toxicodendron carolinianum foliis pinnatis, &c. MAZEAS, *ibid.* x. 595.—Toxicodendron foliis alatis, fructu rhomboideo, DILL. *Elth.* 390, *t.* 292, *f.* 377.—Arbor Americana alatis foliis, succo lacteo, venenato, PLUKENET, *phyt. t.* 145. *f.* 1.

MEDICAL REFERENCES.

DUDLEY, *Phil. trans. abr.* vi. 507.—SHERARD, *ditto.* 508.—KALM, *travels*, i. 77.—MARSHALL, *arbust.* 130.—CUTLER, *Amer. Acad.* 427.—BARTON, *Coll.* 24.—THACHER, *Disp.* 321.—HORSFIELD, *Inaugural Dissertation.*

PLATE X.

Fig. 1. *Rhus vernix, with staminiferous flowers.*

Fig. 2. *A staminiferous or barren flower magnified.*

Fig. 3. *Stamens and rudiment of a pistil.*

Fig. 4. *A fertile flower magnified.*

Fig. 5. *Germ and stigmas.*

Fig. 6. *The fruit.*

Pl. XI.

Conium maculatum

AMERICAN

MEDICAL BOTANY.

CONIUM MACULATUM.

Hemlock.

PLATE XI.

A PLANT bearing the name of *Conium*, κωνειον, has been noted as a poison from remote antiquity. In consequence of the power which it possessed when given in sufficient quantities, of destroying life in a certain and almost immediate manner, it was used at Athens as a mode of execution for those condemned to death by the tribunal of Areopagus. Socrates and Phocion were among the distinguished ancients, who suffered death by the agency of this mortal poison. The accounts which have been left respecting it would lead us to believe that its operation was speedy, and unattended with any violent or long protracted suffering. It was not only employed as an instrument of public executions, but was resorted to by those who sought to encounter suicide in its least formidable shape. Among other instances, that of the Cean old men is related by Ælian, who when they had become useless to the state, and tired of the infirmities of life; invited each other to a banquet, and having crowned themselves as in celebration of a joyous festival; drank the Conium, and terminated their existence together.

The description which has been left by Dioscorides of the Conium, only shews it to have been an umbellate plant, his character of which might apply to many species. The mention made of it by Latin writers under the name of Cicuta are not more satisfactory. Linnæus, influenced by the noxious character of the modern officinal Hemlock, has appropriated to that plant the name of the Grecian species, and most subsequent botanists have followed his example. Haller, however, is of opinion that the ancient poison was not procured from this plant, but from the Cicuta virosa L. a poisonous aquatic, much more powerful and violent in its operation than the common Hemlock. Lamarck adopts the opinion of Linnæus, and believes that the Conium maculatum was really the Grecian poison, and that its properties were rendered more active by the heat of the climate in which it grew. Guersent supposes that the poisonous draught used by the Greeks was not the product of a single species, but a compound prepared from several plants.

Were it not for the tranquillity and ease which attended death from the ancient hemlock, and which Plato has described with interesting minuteness, there would not have been much difficulty in supposing the Grecian plant to be the same with that known at the present day. [*Note* H.] It appears that a large quantity was requisite to insure death. The poison was swallowed in the crude juice, recently expressed from the plant. Of this the draught taken by Phocion was large enough to cost twelve *drachmæ*.* Socrates was prevented from making a libation of a part of the contents of his fatal cup, by being told that the whole was necessary to produce the consummation of his sentence. A large quantity of the modern hemlock might probably have been equally fatal, though with more violent symptoms than those which, if Plato be correct, were experienced by the Athenian philosopher.

The plant, represented in our plate, undoubtedly came to us from Europe. It is now sufficiently common in the United States, about road

* Plutarch, Life of Phocion.

sides and in waste ground, especially in those parts of the country which have been longest settled. It is usually found in bunches, and attains the full height of a man. It flowers from June until the arrival of frost.

The very natural order, called Umbellatæ by Linnæus and Umbelliferæ by Jussieu, to which this plant and the following one belong, is found in the class Pentandria and order Digynia of the Linnæan artificial method.

The genus Conium of Linnæus has *both general and partial involucres, the latter being halved.* The *fruit roundish and furrowed.*

The species maculatum has the *fruit unarmed with the ridges undulated.*

Its more complete description is as follows. Root biennial, somewhat fusiform and generally branched. Stalk round, very smooth, striated, hollow, jointed, and more or less marked with purplish spots. Leaves two or three times pinnate, of a very bright green, with long, sheathing petioles inserted on the joints of the stem ; the leafets pinnatifid and toothed. Flowers in terminal umbels, the general involucre with half a dozen lanceolate, reflected leafets, the partial involucre with three or four situated on the outside. Flowers very small, white. Petals five, oval with their points inflexed. Stamens five, spreading, about the length of the corolla. Germ inferior. Styles two reflexed outwardly. Fruit roundish-oval, compressed, ribbed, the ribs being transversely wrinkled or crenate; separating into two oblong-hemispherical seeds.

Hemlock when fresh has a strong nauseous odour and taste. If the green leaves are distilled, the water which collects in the receiver has an insupportably nauseous taste, while that which remains in the retort is comparatively insipid. This circumstance, and likewise the fact that the dried leaves become inert by age and exposure, render it probable that the chief medicinal efficacy resides in a volatile portion of the plant. Of the more fixed ingredients of Hemlock, a variety of analyses have been made. The most recent which I have met with is that of Schrader, who from a thousand grains of the plant obtained the following substances. Extractive 27.3—Gummy extract 35.2—Resin 1.5—Albumen 3.1—Green fæcula 8.—He also detected various earthy and alkaline salts. These however are found to vary according to the soil in which the plant grows. The volatile portion, which I obtained in water distilled from the leaves, did not exhibit any essential oil, and effected no change in the colour of litmus. It was not altered by sulphate of iron nor acetite of lead.

The Conium maculatum is a narcotic poison, though not of the most powerful kind. Instances of fatal effects from it have been recorded by Dr. Watson in the Philosophical Transactions, and by several other writers. A remarkable case of this sort, which occurred in Spain, is cited by Orfila in his *Traité des poisons.* But there is scarcely any narcotic plant respecting the character of which such various and opposite testimony has been adduced by medical writers. Even the experiments of the same individual are apt to present different results from its use, unless great care be taken in the collection and preparation of the medicine. The truth is, the plant varies exceedingly at different ages, and in different places of growth, and the strength of its preparations is greatly influenced by external circumstances.

When the green leaves of a mature plant which has grown in the sun, or the juice of these leaves, either crude, or properly inspissated, is taken into the stomach; the following symptoms, if the quantity has been sufficient, will rarely fail to take place; viz. a dizziness of the head and nausea of the stomach, a sense of fullness in the eyes and diminished power of vision, together with a general faintness or muscular weakness of the whole body. These sensations usually begin in the course of half an hour. If the dose has been moderate, they will for the most part disappear in the course of half a day, and seldom continue beyond twenty four hours. Larger doses occasion more severe symptoms, as it happens with other narcotics.

The idiosyncrasies of different persons render them variously susceptible of the action of Hemlock. Some are but slightly affected by a quantity, which would prove dangerous to others.

The Hemlock has been for many years a subject of attention with physicians, and has been found a remedy of importance in several diseases. It would occupy a volume to state the whole of the evidences which have been given for and against its use. I shall only mention those complaints in which it has been most employed, and particularly in this country.

In Jaundice.—Dr. Fisher, President of the Massachusetts Medical Society, in his paper on the narcotic vegetables, bears unequivocal testimony in favour of the efficacy of Hemlock in this complaint. He was first induced to employ it with a view to its relaxing effect in facilitating the passage of biliary calculi. Afterwards it was given by him to many icteric patients, and with the exception of three complicated cases, it never failed in his hands or within his knowledge to remove the disease. Dr. Jackson, Professor of the Theory and Practice of Physic in our University, informs me that he has found it of great utility in jaundice, and that except in one or two instances, it has always effected the cure of those cases, which proved susceptible of relief from any medicine. I have repeatedly employed it in the same complaint with indubitable advantage. The dose should be gradually increased until its effects are distinctly felt in the head and stomach. This inconvenience is temporary, and will be preferred by most patients to the evil of a mercurial ptyalism. The yellowness of the skin and eyes, in favourable cases, begins to disappear at an early period, frequently by the second day.—The foregoing practice in jaundice is not new, having been employed in Sweden by Rosenstein, and in other places.

In tic doloureux. In a discourse on this painful disease by Dr. Jackson, published in the New England Journal, Vol. II. a number of cases are detailed, in which perfect relief was afforded by the Hemlock given in large doses, and rapidly increased until a decided effect upon the system was felt. Dr. Jackson recommends to begin with a single grain of the extract, and to increase to five grains for the second or third dose; afterwards to add five grains to every dose until a full effect is felt on the system. In this discourse he cites the experience of Dr. Fothergill, who had employed the Conium successfully in several cases of this disease under a different name. It appears also that some French physicians, whose writings I have not seen, as Chaussier and Duméril, have confirmed the success of our plant in tic doloureux. It must be confessed however, although the Hemlock is more successful in this complaint than perhaps any internal medicine, which has been tried; yet there are cases of such obstinacy, as wholly to baffle the powers of its operation.

In schirrus and cancer. Since the time of Storck, this medicine has been long and abundantly tried, but without any increase of reputation. The experience of modern physicians, and among others of M. Alibert, who tried it in more than a hundred cases in the hospital of St. Louis, have pretty well established the fact, that it is wholly incapable of curing either schirrus or cancer of the confirmed and genuine kind. It is however still administered, rather with a view to its anodyne and palliative effect, than any expectation of radical benefit. In this way its external use is sometimes serviceable.

In old syphilitic affections, it is occasionally useful. It has been recommended in hooping cough, but it is not a perfectly safe medicine for children, owing to the difficulty of ascertaining when its constitutional effects take place in them.—I am informed on the best medical authority, that it is of great use in some cases of hemicrania, which are not regularly intermittent.

The most common form of preserving the Hemlock for use, as well as the most convenient for its exhibition, is that of the inspissated juice or extract. It is well known however, that the extracts kept in our shops differ materially in their strength, so that in beginning from a new parcel, the physician can seldom predict the degree of operation of his first doses. In some instances very great quantities have been taken without the least effect. The extract is apt to prove inert when the plant is gathered too young, when the evaporation is conducted with too much heat, when a decoction of the dried plant has been evaporated instead of the fresh leaves, and lastly when the extract itself has become old. To give the extract its due strength, the plant should be collected at full maturity, while in flower, or in fruit provided it remains green, and the juice or the decoction should be evaporated at the heat of boiling salt water. The stock should be renewed every year. A suitable dose for commencement is from one to five grains. This may be increased at every time of taking it, until its constitutional effects are felt. In beginning the use of a new parcel, more caution is requisite at first, than after its strength has been tested.

The Æthusa cynapium, an umbellate plant very common in Boston, has sometimes been mistaken for Hemlock, which it considerably resembles. It is a smaller plant, with its stalk not spotted. It differs also in having no general involucre, while its partial one is very long.

BOTANICAL REFERENCES.

Conium Maculatum, Linnæus. *Sp. Pl.*—Woodville, *t.* 22.—Curtis, *Fl. Lond.* i. 17.—Smith, *Engl. Bot. t.* 1191.—Pursh, i. 195.—Cicuta vulgaris, Morison, *Umb. t.* 6.—Parkinson, *Theatr.* 933.—Cicuta major, Lamarck, *Encyclopedie Methodique.*—Cicutaria vulgaris, Clusius, *Hist.* 200.

MEDICAL REFERENCES.

Murray, *Apparatus medicaminum,* i. 322.—Cullen, *Mat. Med.* ii. 263.—Fothergill, *Med. Obs.* iii. 400.—Hunter *on the venereal,* 108, 175, 199, 247, &c.—Home, *Annals Med.* iii. 66.—Butter, *Med. Comment.* i. 373.—Fisher, *Mem. Mas. Soc.* i.—Jackson, *N. Engl. Journ.* ii. 105.—Guersent. *Dict. Sciences Medicales,* v. 208.—Orfila, *Toxicologie,* iii. 279, &c. &c. &c.

PLATE XI.

Fig. 1. *A branch of Conium maculatum.*

Fig. 2. *Flower magnified.*

Fig. 3. *Fruit magnified.*

Pl. XII.

Cicuta maculata

CICUTA MACULATA.

American Hemlock.

PLATE XII.

It is a rule sanctioned by the observations of medical botanists, that umbelliferous plants, which grow in or about the water, are of a poisonous nature. This rule will generally be found correct, although it has exceptions. As far as aquatic plants of this natural order have been examined, their properties, in a great majority of instances, have been found, more or less of a deleterious kind. The *Cicuta virosa* of Europe is a highly poisonous plant, possessing such formidable activity that its internal use is hardly attempted in medicine. An American species, the *Cicuta maculata*, the subject of this article, is very closely allied in its botanical habit to the European plant, and was equally deserving of suspicion from its appearance, although the public were not generally aware of its true character. Within a few years past, several instances have been brought to light of fatal effects ensuing from this plant being incautiously eaten by children. It is therefore necessary that the species should be suitably designated, that a source of so much danger may be known and avoided.

The Cicuta maculata, to which I have applied the name of *American Hemlock*, not having heard any common appellation except that of *Snakeweed*, inhabits wet meadows and banks, from the northern to the southern limits of the United States, flowering in July and August. It is so frequently cut with hay, among which it often grows in large quantities, that we might expect to see its deleterious properties operating on domestic cattle, were it not that their bodies are probably less susceptible of its poison than ours. The European Cicuta, above mentioned, is highly noxious to man, and to some domestic animals, yet goats and sheep eat it with impunity.

The genus *Cicuta* differs from other genera of umbellate plants in having *no general involucre*, a short, partial involucre, and a *fruit* which is nearly *orbicular, compressed* and *furrowed.**

* This description of the fruit agrees with the present species and also with *Cicuta bulbifera*, a smaller species not uncommon about Boston. The Cicuta virosa of Europe I have never seen.

The species maculata has a fascicled root and oblong leaves with mucronate serratures.

The class and orders are as in the last article.

This plant is so remarkable for the form of its root, that had not the name of *maculata* been confirmed to me by the best authorities, I should have thought that of *fasciculata* to be greatly preferable. This root is composed of a number of large, oblong, fleshy tubers, diverging from the base of the stem, and frequently being found of the size and length of the finger. The root is perennial, and has a strong, penetrating smell and taste. In various parts of the bark it contains distinct cells or cavities, which are filled with a yellowish resinous juice. The plant is from three to six feet high. Its stem is smooth, branched at top, hollow, jointed, striated, and commonly of a purple colour, except when the plant grows in the shade, in which case it is green. The leaves are compound, the largest being about three times pinnate, the uppermost only ternate. Most of the petioles are furnished with long obtuse stipules, which clasp the stem with their base. Leafets oblong acuminate, serrate, the serratures very acute or mucronated. The veins end in the notches, and not at the points of the serratures. The flowers grow in umbels of a middling size, without a general invo-

17

lucre. The partial umbels are furnished with involucres of very short, narrow, acute leafets. The distinctness or separation of these umbels characterizes this plant at a distance among other plants of its kind, whose umbels are more crowded. Calyx of five very minute segments. Petals five, white, obovate with inflected points. Fruit nearly orbicular, compressed, ten furrowed, crowned at top, and separating into two semicircular seeds.

The fleshy root of the Cicuta maculata, when pressed, emits from its divided extremities a viscid yellowish juice of a strong penetrating taste. This juice dissolves in alcohol, from which it is precipitated by water. When distilled, a thick volatile oil collects in the receiver in the form of a film upon the surface of the water. The remainder of the juice yields a resin of a dark orange colour, fusible and inflammable. The decoction of the root affords a pearl coloured fluid, not very sensible to the tests of mucus, fæcula, tannin or extractive.

In August 1814, an account was sent to Boston by Dr. Stockbridge of Bath (Maine) of the effect produced on three boys by eating a poisonous root, which they had dug up, supposing it to belong to the plant called "Life of man." One of them was siezed with violent convulsions, frothed at the mouth, and died in an hour and a half. The other two were affected with vomiting, stupor, dilatation of the pupil, great paleness and universal distress; which symptoms disappeared in one in twenty four, and in the other in thirty six hours. It was supposed that the first boy had swallowed about a drachm of the root, and the others about half that quantity. A specimen of the plant was sent to me at the same time with the account, and proved to be the Cicuta maculata. Dr. Stockbridge's letter, which was published in the New England Journal, contains two other cases of the effect of this root, in one of which it proved fatal.

Shortly after the publication of the above facts, an article appeared in the New York Medical Repository, containing an account by Dr. Ely of Dutchess county, of the effects of an unknown poisonous root, supposed to be the white hellebore. Three small boys, who had gone into a meadow in search of sweet flag root, had dug up and eaten another root by mistake. Two of them died in convulsions in about an hour after they had swallowed it. They discharged much blood and froth from the mouth and nose; their eyes were fixed, with the pupils dilated, and a rapid motion of the eye lids. The third boy vomited, and recovered. When taken to the place the next day, he pointed out the spot where they had dug the root, and where a considerable quantity of it remained. Some of the root was planted by Dr. Mitchill in the New York Hospital garden, where it vegetated and produced flowers and fruit. It turned out to be the Cicuta maculata of Linnæus. In the same article, is a letter from Dr. Muhlenberg, stating that he had received specimens from Savannah and from West Pennsylvania, where it had destroyed several persons, who ate it by mistake for angelica. All the specimens were similar, so that there could be no doubt of the identity of the plant. In the same letter, Dr. Muhlenberg remarks, that he had reason to believe that the poisonous quality of the root is altered by cultivation in a dry soil.

The foregoing facts are sufficient to establish the poisonous character of the plant under consideration. They may also serve to shew the importance of accurate descriptions and faithful engravings of noxious vegetables, which may enable even unlearned observers to distinguish them at sight. There can be little doubt that cases, like those above described, have occurred in repeated instances, which have never met the public eye. Perhaps also from an ignorance of the real cause of the symptoms, the proper remedies have been neglected. The plant is extremely common in many parts of the United States, and I believe its true character is not generally suspected. A very respectable physician informed me, that it was used in his vicinity as a gargle for sore throats, by people unsuspicious of its qualities.

Since the discovery of its narcotic properties, the Cicuta has been used in small doses, as a substitute for the conium, by one or two practitioners in this place. Its effects were very analogous to those of the true hemlock, as far as they were observed, but more powerful. A primary symptom, which attended a large dose, was nausea and vomiting.

The treatment of persons poisoned by this plant, as in the case of other narcotics, should primarily consist in a thorough evacuation of the stomach. As there commonly exists a spontaneous tendency to vomit, occasioned by the poison itself, this should be assisted by mechanical means, by irritating the throat with the finger, or with a feather. Of emetics, the sulphate of zinc is to be preferred, on account of its speedy operation. Castor oil or infusion of senna, should be given as soon as vomiting has taken place. The vegetable acids, such as lemon juice or vinegar, have a neutralizing influence on the narcotic, and are therefore useful. Strong coffee and tea are the best antidotes for the stupor, and should be promptly administered. In violent cases, bloodletting should be resorted to. As most narcotic poisons act by destroying the functions of the brain, respiration being suspended, because it is under the influence of that organ; Mr. Brodie is of opinion, that in some cases, life might be preserved by keeping up artificial respiration, after death has apparently taken place.

BOTANICAL REFERENCES.

Cicuta maculata. LINNÆUS *Sp. pl.*—PURSH, i 195.—Ægopodium foliis lanceolatis, acuminatis, serratis. GRONOVIUS, *Virg.* 32.—Angelica Caribæarum elatior, olusatri folio; flore albo; seminibus luteis, striatis, cumini odore et sapore? PLUKENET, *Alm.* 31, *Phyt. t.* 76, *f.* 1.

MEDICAL REFERENCES.

SCHŒPF, 36.—BART. *Coll.* 18, 46.—STOCKBRIDGE, *New Engl. Journal*, iii. 334. MITCHILL, ELY and MUHLENBERG, *Med. Repository*, xvii. 303.

III.

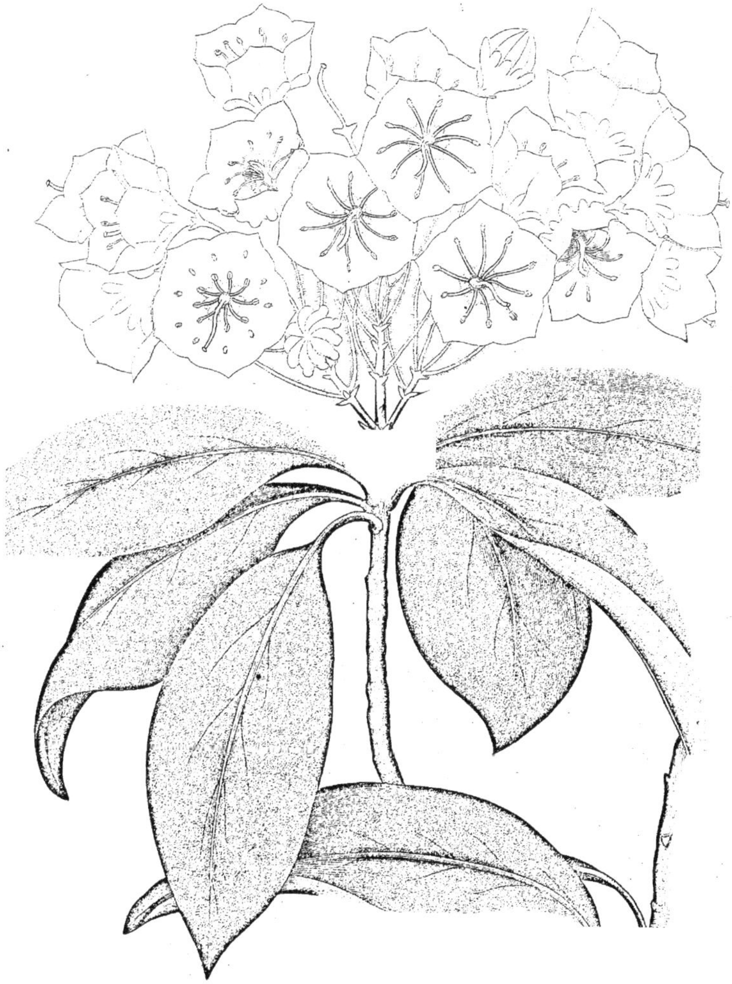

KALMIA LATIFOLIA.

Mountain laurel.

PLATE XIII.

THE Swedish botanist, Peter Kalm, a pupil of Linnæus, who travelled in North America in 1748—9, has had the honor of giving name to one of the most elegant genera of flowering shrubs which our continent produces. The genus named *Kalmia* by Linnæus, includes several species, of singular beauty, among which the Mountain laurel is much the largest and most elegant, as well as the one whose properties have received most attention. Its occurrence in the United States is very frequent, and its common appellations of course various. The names of *Laurel*, *Lambkill*, *Ivy*, *Spoonwood*, and *Calico bush*, it seems, are applied to it in various parts of the country. This shrub grows in the southern parts of New Hampshire, and is occasionally met with throughout Massachusetts. In the Middle States it becomes more frequent, and it is said to extend near to the southern limits of the Union. Michaux, in his account of the forest trees, states, that it is particularly abundant through the whole range of Allegany mountains, upon the borders and near the sources of rivers. It gradually diminishes however on both sides as these rivers approach to the sea, or to their confluence with the great western streams.

The botanical character of the genus consists *in a five parted calyx, a hypocrateriform corolla, containing ten depressions in its border, in which the anthers are lodged; a capsule five celled.*

The specific character is, that the *leaves are scattered, petioled, oval and smooth; the corymbs terminal, viscid and pubescent.*

Class *Decandria*, order *Monogynia*. Natural orders *Bicornes*, Linn. *Rhododendra*, Juss.

The height of the Kalmia latifolia is generally that of a shrub, sometimes however attaining to the altitude of a small tree. Its leaves are evergreen, coriaceous, very smooth, with the under side somewhat palest. Their form is oval, acute and entire; their insertion by scattered petioles, on the sides and extremities of the branches. The flowers vary from white to red; they grow in terminal corymbs, simple or compound with opposite branches, and made up of slender peduncles. These are invested with a glutinous pubescence, and supported at base by ovate, acuminate bractes. The calyx is small, five parted, persistent, with oval, acute segments. The corolla is monopetalous, with a cylindrical tube, a spreading disc, and an erect, five cleft margin. At the circumference of the disc on the inside, are ten depressions or pits, accompanied with corresponding prominences on the outside. In these depressions the anthers are found lodged at the time when the flower expands. The stamens originate from the base of the corolla, and bend outwardly, so as to lodge their anthers in the cells of the corolla. From this confinement they liberate themselves during the period of flowering and strike against the sides of the stigma. The germ is roundish, the style longer than the corolla and declined, the stigma obtuse. Capsule roundish, depressed, five celled and five valved, with numerous small seeds.

I have examined chemically the leaves of the Kalmia, gathered at the time the shrub was in fruit. The following constituent principles were found to exist in them.

1. Vegetable mucus. This exists in large quantities, and is dissolved in water both by infusion and decoction, rendering it extremely mucilaginous or ropy. When alcohol is added to this se-

lution, the mucus separates in the form of a flocculent coagulum, which is tough and stringy, and on drying has a brownish colour. When chewed, it soon fills the mouth with mucilage.

Silicated potash rendered the upper stratum of the liquid dark and opaque, but without any precipitate like that which takes place in the mucilage of gum.

2. Tannin. This is readily thrown down from the decoction and tincture by gelatin. The sulphate of iron strikes with it a very black colour.

3. Resin. This also exists plentifully. It communicates to alcohol a reddish colour, and is instantly precipitated from it by water. When obtained pure, it is of a reddish cast, fusible, inflammable and moderately bitter.

I have not detected any extractive, properly so called, in these leaves. When the muriate of tin is added to the decoction, it separates a very copious yellow precipitate. This however is owing to the mucus. If alcohol be first added to the decoction, and the coagulum which it forms withdrawn ; the fluid no longer gives a precipitate with muriate of tin, although it readily yields one to gelatin.

Distillation with water affords a mild fluid with little taste or odour.

The Kalmia latifolia, together with some other species of its genus, has long had the reputation, in various parts of the country, of being poisonous to certain domestic animals. Catesby says of it, that "deer feed on its green leaves with impunity; yet when cattle and sheep, by severe winters deprived of better food, feed on the leaves of this plant, a great many of them die annually."

Kalm, the Swedish traveller, who gave name to this genus, says of Kalmia latifolia, "The leaves are poison to some animals, and food for others; experience has taught the people, that when sheep eat of these leaves, they either die immediately, or fall very sick, and recover with great difficulty. The young and more tender sheep are killed by a small portion, but the older ones can bear a stronger dose. Yet this food will likewise prove mortal to them, if they take too much of it. The same noxious effect it shews in regard to calves which eat too much of the leaves; they either die, or do not recover easily. I can remember that in the year 1748 some calves ate of the leaves; they fell very sick, swelled, foamed at the mouth and could hardly stand; however, they were cured by giving them gunpowder and other medicines. The sheep are most exposed to be tempted with these leaves in winter, for after having been kept in stables for some months, they are greedy of all greens, especially if the snow still lies upon the ground, and therefore the green but poisonous leaves of the Kalmia are to them very tempting. Horses, oxen and cows, which have eaten them, have likewise been very ill after the meal, and though none of them ever died of eating these leaves, yet most people believed, that if they took too great a portion of them, death would certainly be the result." "On the other hand, the leaves of the Kalmia are the food of stags, when the snow covers the ground and hides all other provisions from them. Therefore, if they be shot in winter, their bowels are found filled with these leaves, and it is very extraordinary, that if those bowels are given to dogs, they become quite stupid, and, as it were, intoxicated, and often fall so sick, that they seem to be at the point of death; but the people who have eaten the venison have not felt the least indisposition."—*Travels in North America*, vol. i.

There is a common belief, that the flesh of the American Pheasant or Partridge is at certain times imbued with a poisonous quality. This circumstance has been attributed (I know not with what evidence) to their feeding in winter upon the buds of the Kalmia. Mr. Wilson, the ornithologist, informs us, that he has sometimes found the crops of these birds distended almost entirely with laurel buds; but that he has eaten freely of the flesh of these very birds, without any ill consequence whatever.

On the human system, the Kalmia has been also said to manifest a deleterious influence. The late Professor Barton has adduced some evidences of its noxious character.* He states that the Indians make use of a decoction of the leaves to destroy themselves. In an Inaugural Dissertation on two species of Kalmia, the latifolia and angustifolia, by Dr. G. K. Thomas, we are told that the leaves of these shrubs possess a decidedly narcotic property. I have not recently seen Dr. Thomas' Dissertation, and therefore quote from memory and from extracts. From his experiments however it appeared, that a very small quantity was sufficient to produce sensible inconvenience. Thirty drops of a strong decoction, given six times a day, are said to have occasioned so much vertigo, as to render it necessary to diminish the frequency of its exhibition.

From my own experience, I am not disposed to think very highly of the narcotic power of the

* Dr. Barton states, that a few drops of the tincture poured upon the body of a large and vigorous rattlesnake, killed the reptile in a very short time.

Kalmia. I have repeatedly chewed and swallowed a green leaf of the largest size, without perceiving the least effect in consequence. I have also seen the powder, freshly made from leaves recently dried, taken in doses of from ten to twenty grains, without any subsequent inconvenience or perceptible effect. The taste of these leaves is perfectly mild and mucilaginous, being less disagreeable than that of most of our common forest leaves.

I am inclined to believe that the noxious effect of the Kalmia upon young grazing animals may be in some measure attributed to its indigestible quality, owing to the quantity of resin contained in the leaves.

An ointment made of the powdered leaves has been recommended in tinea capitis and some other cutaneous affections. I have seen an eruption, very much resembling psora, removed by it.

The wood of the Kalmia is hard and dense, approaching in its character to that of box. It is much used for the handles of mechanics' tools, &c. and it has even been employed as a material for musical instruments. As an ornamental shrub, this species stands in the highest rank, and by the frequency of its growth and the brilliancy of its flowers, it contributes in a great degree to the elegance of the natural scenery in those mountains and woods, which it inhabits. When cultivated in gardens, it requires a soil which is somewhat moist, and a shady or northern aspect.

BOTANICAL REFERENCES.

Kalmia latifolia, Lin. *Sp. pl.*—Curtis, *Bot. Mag. t.* 175.—Michaux *f. Arbres forestiers*, iii. 147, *t.* 5.—Pursh i. 296.—Chamædaphne foliis tini, &c.—Catesby, *Carolina*, ii. *t.* 98.—Ledum floribus bullatis. &c. Trew, *t.* 38.—Cistus chamærhododendros, &c. Plukenet, *Phyt. t.* 379, *f.* 6.

MEDICAL REFERENCES.

Kalm, *travels*, i. 335, &c.—Bart. *Coll.* i. 18, 48; ii. 26.—Thacher, *Disp.* 247.—Thomas, *Inaugural dissertation.*

PLATE XIII.

Fig. 1. *Branches of Kalmia latifolia with flowers and fruit.*

Fig. 2. *Stamens.*

Fig. 3. *Calyx and pistil.*

SPIGELIA MARILANDICA.

Carolina Pink root.

PLATE XIV.

We are told by different writers, that this fine plant is a native of all the southern states from Pennsylvania to Georgia and Louisiana, growing in rich soils, especially about the borders of woods. It does not bear the severity of a northern winter. For my living specimens I was indebted to my excellent and learned friend, the late Dr. James Macbride, of Charleston, S. C.

The genus Spigelia has *a funnel shaped corolla* and *a capsule, which is double, two celled and many seeded.* The species Marilandica is *perennial, with a simple stem and opposite leaves.*

Class *Pentandria;* order *Monogynia.* Natural orders *Stellatæ*, Lin. *Gentianæ*, Juss.

The root of the Spigelia Marilandica is perennial, with many fibrous branches. The stalks proceed several from a root; they are simple

Pl.XIV.

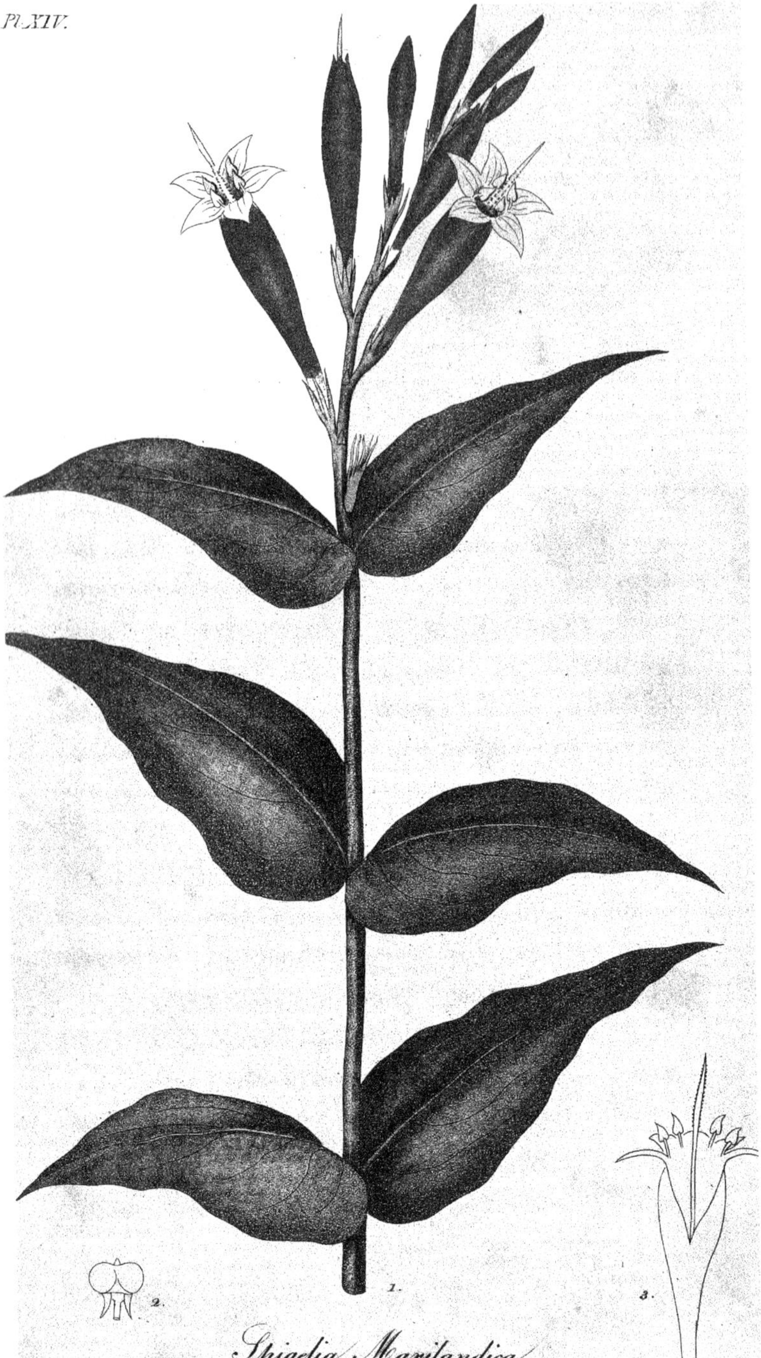

Spigelia Marilandica

four sided and nearly smooth. Leaves opposite, sessile, ovate, acuminate, entire, smooth, with the margins and veins sometimes pubescent. The stalk commonly terminates in a simple one-sided raceme of flowers, although I have seen luxuriant specimens with two. The peduncles are extremely short, so that the raceme may without impropriety be denominated a spike. Calyx persistent, with five linear-subulate, finely serrulate leaves, which are reflexed in the ripe fruit. Corolla five times as long as the calyx, scarlet or crimson without, orange coloured within, the tube inflated and angular at top, the border divided into five acute, spreading segments. Stamens very short, inserted into the mouth of the corolla between the segments ; anthers oblong-heart shaped. Germ small, superior, ovate. Style longer than the corolla, jointed near its base and bearded at the extremity. Capsule double, consisting of two, cohering, one celled, globular portions, seated on a common receptacle.

The Spigelia is a mucilaginous plant, with a mild and not very disagreeable taste. The infusion and decoction of the root and leaves afford a flocculent precipitate with alcohol. They are discoloured but not precipitated by silicated potash. They have little sensibility to gelatin, although the tincture is made turbid by it. After the decoction was filtrated from the mucus, which had been coagulated by alcohol, it gave a precipitate with nitrate of mercury, but none with muriate of tin. Sulphate of iron caused a dark green precipitate from the decoction, and but little change in the tincture. No distinct evidence of resin presented itself. A substance which may perhaps be considered a variety of extractive matter, appears to exist in this plant, as the tincture was affected in nearly the same manner by the salts of tin and mercury above mentioned, as the filtrated decoction.

Water may be considered an adequate solvent for the chief proximate principles of this plant.

The medicinal reputation of the Spigelia is founded on the powers which it is supposed to possess as a vermifuge. This reputation is now so generally established, that the plant has become a considerable article of commerce to various parts of the world, from our southern states. This is a sufficient evidence, that the medicine has, to a certain extent, satisfied public expectation, and obtained the sanction of practitioners. But beyond this, it is difficult to speak confidently on the subject. The Spigelia belongs to a class of medicines, which are frequently prescribed, without positive proof of the existence of the cause which they are intended to remove; which often fail altogether in the hands of the most successful practitioners; which frequently succeed merely because they are backed with medicines of a more active class; and whose apparent success is sometimes the consequence solely of a diseased state of the body.* Our plant is however entitled to trial, especially where it can be obtained fresh, and in full strength. A physician of the southern states, for whose opinion I have much respect, Dr. Norcom of Edenton, N. C. informed me some years since, that the Spigelia was

* From the list of equivocal anthelmintics, I would except those which have a cathartic operation, also a number of mineral origin. But I am fully persuaded, that many reputed vermifuges have enjoyed a reputation which they do not deserve. The *Dolichos pruriens* has received the commendations of practitioners and medical writers, on the presumption that its spiculæ exert the same stimulant effect on the bodies of worms in the alimentary canal, that they do on the human skin externally. I was long ago inclined to doubt the power of these spiculæ to withstand the digestive process of the stomach. My suspicions were confirmed upon finding that simple maceration in warm water for an nour, dissolves their virus, and renders them incapable of producing their usual stimulus of itching, when applied to the skin. Some late experiments by my pupil, Dr. Chandler, have shewn that the gastric juice destroys their activity in the same manner.

It is not necessary in this place to revert to the *Fern root* of Madame Nouffer, and various other exploded anthelmintics of its kind.

most active when recently dried, and that its efficacy was always impaired by keeping more than six months. Dr. Garden had previously made observations somewhat similar. If this be the case, we may account for its failures in the hands of those who obtain it at a distance when half a dozen years old.

Drs. Lining, Garden, and Chalmers of Carolina, are the writers who first introduced the Spigelia to notice, and who have spoken most unequivocally in its praise. Each of these physicians has represented it as an anthelmintic of superior efficacy. It appears that under certain circumstances, it is capable of operating as a cathartic, and that in these instances, the most advantage has been experienced from it. Dr. Garden says, that he had given it in hundreds of cases, and that he "never found it do much good except when it proved gently purgative." As the action of the Spigelia upon the bowels is quite uncertain, most practitioners either unite, or follow it with calomel or some purgative medicine.

We are told that the pink root, when in its most active state, if given in large quantities, induces narcotic symptoms, such as stupor, headach, dilated pupil, &c. Dr. H. Thompson, who took large doses of the root to try its effect on himself, found that it produced an increased quickness of the pulse, drowsiness, flushing of the face and stiffness of the eyelids. Dr. Chalmers attributes to its too free use the cases of two children, who died in convulsions. Dr. Macbride informs us that its narcotic effects are seldom or never attended with danger, and that some physicians consider them an evidence of the favorable operation of the medicine. The opinion that this effect is owing to the root of some deleterious plant taken up with the Spigelia, seems to be void of foundation.

As in most other perennial plants, the root of the Spigelia possesses a greater share of activity than the herb. Of this root ten grains may be given in powder to a child four years old, twenty to one which is seven, and a drachm to an adult. If no inconvenience ensue, it may be repeated two or three times a day. If the infusion is preferred, an ounce of the root may be infused in a pint of water, and half the quantity taken by an adult or one or two spoonfuls by a child.

BOTANICAL REFERENCES.

Spigelia Marilandica, LINN. *Sp. pl.*—CURTIS, *Bot. Mag. t.* 80.—WOODVILLE, ii. *t.* 105.—WALTER, *Flor. Car.* 92.—MICHAUX, i. 147.—PURSH, i. 139.—ELLIOTT, i. 236.—Lonicera spicis terminalibus, &c. GRONOV. *Virg.* 30.—Periclymeni Virginiani flore coccineo planta Marilandica, &c. CATESBY, ii. *t.* 78.

MEDICAL REFERENCES.

CHALMERS, *on the weather and diseases of South Carolina,* i. 67.—LINING, *Essays, phys. and lit.* i. 436. GARDEN, *ditto,* iii. 145.—HOME, *Clin. exper.* 420.—MURRAY, *App. Med.* i. 548—MACBRIDE, *in Elliott's Car.* 237.—THOMPSON, *Inaug. Diss.*

PLATE XIV.

Fig. 1. *Spigelia Marilandica.*

Fig. 2. *The capsule with the reflected calyx.*

Fig. 3. *Corolla opened.*

Pl.XV.

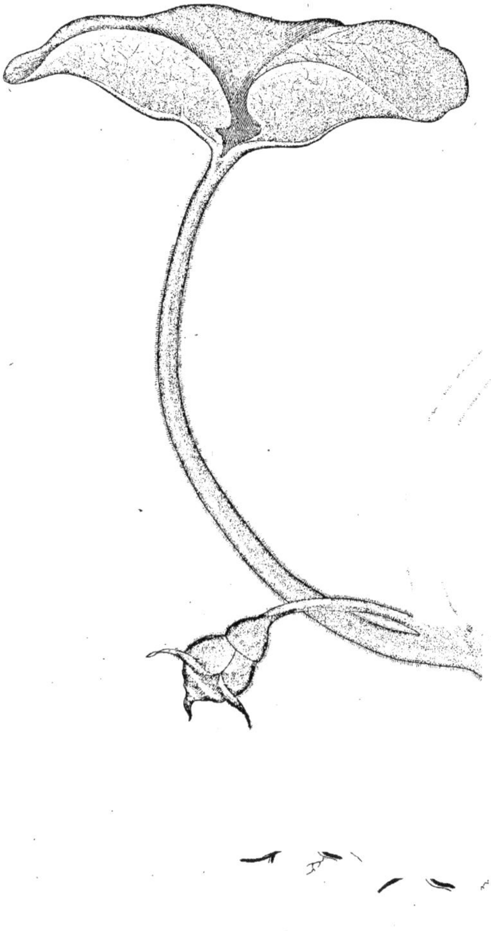

Asarum Cana

ASARUM CANADENSE.

Canada Snake root. Wild ginger.

PLATE XV.

THE properties of this mild aromatic have been so far misconceived, probably from its relation to an European species, that it would be improper in a work of this kind, to pass over it without notice of its real character. It affords a striking exception to the rule, that botanical affinities are capable of indicating the medicinal qualities of vegetables. This plant in its effect on the human system, is widely different from the European asarabacca, although it approaches it so nearly in its form, that Michaux styles it "vix distinctum ab Europæo."

The Asarum Canadense grows in old woods and mountainous tracts from Canada to Carolina. It is one of the humblest plants, presenting only two leaves with their stalks, which appear to constitute the whole of the plant above the ground.

On plucking the plant, the two leaves are found connected below, with an obscure flower in their fork, which had rested on the surface of the ground, or been buried under the decayed leaves and soil. Its flowering time is from May to July.

This plant, from the number of its stamens, is placed by Linnæus and Michaux in the class *Dodecandria.* Pursh, who has omitted this class in his Flora, has transferred the Asarums to Gynandria, from the circumstance that the stamens are inserted on the germ. This place however is not better suited to the Asarum, than to a multitude of other plants whose germ is inferior.

Linnæus' natural order for this plant is *Sarmentaceæ* and Jussieu's *Aristolochiæ.*

Generic character. *Calyx three or four cleft, superior; corolla none; anthers growing to the middle of the filaments. Capsule coriaceous, crowned.*

Specific character. *Leaves two, reniform; calyx woolly, cleft to the base; its segments spreading at top.*

The root of the Asarum is creeping, fleshy, and somewhat jointed. Leaves kidney shaped, pubescent on both sides, with long, round, hairy petioles. Flower solitary, growing from the fork of the stem, on a pendulous hairy peduncle. Calyx very hairy or woolly, consisting of three broad, concave leafets, which are mostly of a brownish or dull purple on the inside at top and bottom, and terminated by a long, spreading, inflected point, with reflexed sides. The colour varies greatly according to the amount of light which the plant enjoys, being sometimes nearly green. Stamens twelve, inserted on the germ at a distance from the calyx, the alternate ones longer. Anthers growing to the filaments below their extremity. Near the divisions of the calyx are three short, curved, filamentary substances, which may perhaps be called nectaries. Germ inferior, somewhat hexagonal, marked at top inside with a dark red line; style conical, striate, parted at top into six recurved, radiating stigmas.

The root of the Asarum has an agreeable aromatic taste, which is intermediate between that of ginger and the aristolochia serpentaria. This quality has given it the names of *Wild ginger* and *Snake root* in different sections of the country. The name *Colt's foot* is also applied to it.

The chemical trials, to which I have subjected the root, bring to view the following substances:— 1. A light coloured, pungent, volatile oil, possessing the characteristic taste and smell of the plant in a high degree. 2. A resin, which is of a reddish colour and very bitter. These two constituents communicate to alcohol the active properties of the plant. 3. Fœcula. 4. A gummy mucus. These exist in such quantities as to impede the filtration of the decoction. Astringency hardly exists in this root, as a gelatinous solution gave no evidence of tannin, and the sulphate of iron produced a green colour hardly bordering on black.

It has been asserted, and the statement copied from one book to another, that the Asarum Canadense is a powerful emetic. I presume that subsequent writers have taken their opinion from Cornutus, who, in his plants of Canada, informs us, that two spoonfuls of the juice of the leaves of the Asarum, (meaning the European plant, rather than the American,) are found to evacuate the stomach powerfully. I can hardly doubt, that if such an operation has really been produced from the Canadian species, it must have taken place in irritable stomachs, to whom two spoonfuls of any crude vegetable juice would have proved emetic. Having seen the root of this plant used in the country in considerable quantities as a sudorific, I was long since induced to doubt its emetic power. Subsequent experience has satisfied me that the freshly powdered root, given to the extent of half a drachm, and probably in still larger quantity, excites no vomiting nor even nausea.

Still however the plant deserves not to be discarded from use. The aromatic flavour of the root is more agreeable than that of the aristolochia serpentaria, which article it seems to resemble in its medicinal powers. Several country practitioners, who have employed it, have spoken to me favourably of its effect, as a warm stimulant and diaphoretic. As a substitute for ginger, in common domestic use, I know of no indigenous article which promises so fairly as this.

Alcohol is the proper solvent for the active properties of this plant. The tincture has a dark red colour, and a highly concentrated taste of the root.

BOTANICAL REFERENCES.

Asarum Canadense, LIN. *Sp. pl.*—MICHAUX, i. 279.—PURSH, ii. 596.—Asarum foliis reniformibus, mucronatis, binis, GRONOVIUS, 72.—Asaron Canadense, CORNUTUS, *Canad.* 24, *t.* 25.—Asaron Americanum, PARKINSON, *theatr.* 266.

MEDICAL REFERENCES.

SCHŒPF, 72.—BART. *coll.* 26, 48.—COXE, *Disp.* 368.

PLATE XV.

Fig. 1. *Asarum Canadense.*

Fig. 2. *The germ with the stamens and nectareous filaments.*

Fig. 3. *A petal.*

Fig. 4. *A stamen a little magnified.*

Fig. 5. *Style and stigmas magnified.*

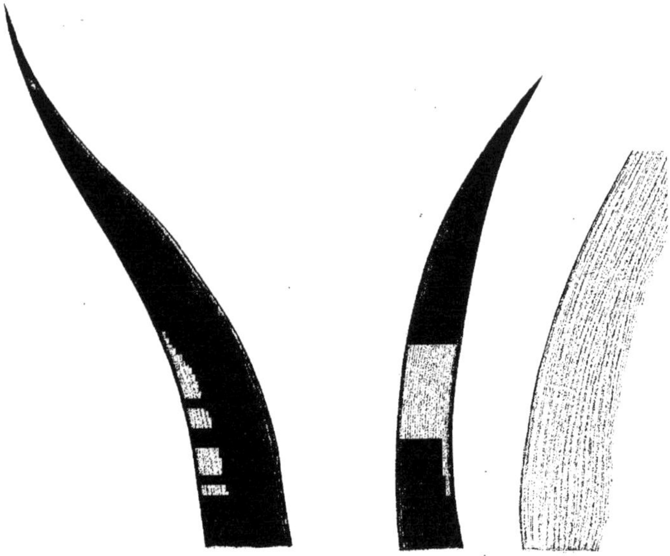

Iris versicolor

IRIS VERSICOLOR.

Blue Flag, or Flower de luce.

PLATE XVI.

In the Hortus Elthamensis, published by John J. Dillenius in 1732, there are figures of two plants under the name of "Iris Americana versicolor," the one with an entire, the other with a crenate style. To one of these, the plant represented in our plate apparently belongs. This plant however is so subject to variation in the colour of its flowers, the crenatures and direction of its stigmas, &c. that it has received from different botanists dissimilar names. The Linnæan characters of Iris versicolor and Virginica are hardly sufficient to distinguish them from each other. Our plant is the Iris versicolor of Muhlenburgh's catalogue, by his own declaration. In the character of its stem however, it agrees equally well with Iris virginica of Linnæus and Michaux. It may be doubted whether the plant figured in the Botanical Magazine, *t.* 703, is more than a variety of this species. The characters taken from the comparative length of the stem and leaves, of the inner petals and stigmas, and the direction of the stem and of the stigmas ; are all subject to variation. Michaux, Elliott and Pursh make the Virginica synonymous with Iris hexagona of Walter, which seems permanently distinguished by the deep furrows in the angles of its capsule.

The Iris versicolor is found throughout the United States in the borders of swamps and in wet meadows, of which it forms a principal ornament in the month of June. No race of vegetables can be better marked than the elegant genus to which this plant belongs. They are essentially distinguished by *a corolla, parted into six segments or petals, of which three are reflexed and three are erect. The stigmas resembling petals.* The species in our plate has *ensiform leaves, its stem acute on one side, its capsules oblong, three sided, with obtuse angles.*

Class *Triandria,*—Order *Monogynia.*—Natural orders, *Ensatæ,* Lin. *Irides,* Juss.

The root is fleshy, horizontal, sending down a multitude of fibres. Stem two or three feet high, round on one side, acute on the other, frequently branched, and bearing from two to six flowers.

Leaves sword shaped, striated, sheathing at base. Bractes becoming scarious. Peduncles of various length, flattened on the inside. Germ three cornered, with flat sides and obtuse angles. Outer petals of the flower spatulate, beardless, the border purple, the claw variegated with green, yellow and white, and veined with purple. Inner petals erect, varying in shape from spatulate to lanceolate, usually paler than the outer, entire or emarginate. Style short, concealed; stigmas three, petal-form, purple or violet, resting on the outer petals, their extremeties bifid, crenate, and more or less reflexed; their lower lip short. Stamens concealed under the stigmas, with oblong-linear anthers. Capsule three celled, three valved; when ripe, oblong, turgid, three sided, with roundish angles. Seeds numerous, flat.

The young leaves of this and some other species of Iris, afford an excellent view of the spiral filament, which lines the sap vessels of the leaf. If a leaf, which has just emerged from the ground, be carefully broken across, and the segments gradually drawn asunder, these fine filaments will unroll themselves, and their spiral structure become very obvious to the microscope.

The root of the Iris versicolor has a nauseous taste, and when swallowed or held in the mouth.

even in small quantities, it leaves behind a powerful sense of heat and acrimony in the fauces. Its most active chemical constituent appears to be a resin, which separates in the form of a white precipitate, when water is added to the alcoholic solution. The decoction suffers little or no change with alcohol, gelatin or salts of iron. Muriate of tin affects it slightly, the nitrate of mercury more abundantly. Its taste is much weaker than that of the tincture. Water distilled from the root has a highly nauseous taste and odour.

The root of the Iris versicolor given medicinally is an active cathartic. Mr. William Bartram, in his travels in Georgia and Florida, informs us, that on his arrival at Ottasse, an Indian town on the Tallapoose, he found the natives "fasting, taking medicine, and praying, to avert a grevious calamity of sickness which had lately afflicted them, and laid in the grave abundance of their citizens. The first seven or eight days, during which time they eat or drink nothing, but a meagre gruel made of a little corn-flour and water; taking at the same time, by way of medicine or physic, a strong decoction of the roots of the *Iris versicolor*, which is a powerful cathartic. They hold this root in high estimation, and every town cultivates a little plantation of it, having a large artificial pond just without the town, planted and almost overgrown with it."

Having myself formerly made use of this root in dispensary practice, I can bear testimony to its efficacy as a medicine, though not altogether to its convenience. A small quantity of the recent root, or a few grains of the root newly dried, are generally certain and active in their operation on the bowels. They are however apt to occasion a distressing nausea like sea sickness, with a prostration of strength of some hours' continuance; so that I think the plant will not be like to come into favor as a cathartic, at least when better ones are at hand. The activity of this article is diminished by age.

The stimulating properties of the Iris render it capable of exciting many of the secretions, as well as excretions. But I know of no purpose for which it seems better calculated, than that of a diuretic. The late Dr. Macbride of Carolina assured me, that he had found great benefit in dropsical affections from a decoction of the roots of this plant in combination with those of Eryngium yuccifolium. In consequence of his recommendation, I administered the tincture of the Iris in small doses to several persons affected with anasarca and with hydrothorax. It was evidently of service to a majority of those who took it, for a certain time. That it did not always cure the disease, is a reproach which it must divide with diuretic remedies of much older celebrity.

The Iris gracilis, a species described in the Florula Bostoniensis, the Iris pseudacorus of Europe, and several others of the genus, appear to possess properties very similar to those of the plant described.

BOTANICAL REFERENCES.

Iris versicolor, LIN. *Sp. pl.*—DILLENIUS, *Elth. t.* 155.—CURTIS, *Bot. Mag. t.* 21, *a variety.*—PURSH, i. 29.—ELLIOTT, *Car.* i. 45. WALTER, *Car.* 67.

MEDICAL REFERENCES.

BARTRAM, *travels*, 454, *Lond. edit.*—CUTLER, *Mem. Amer. Acad.* 405—6.—MACBRIDE, *in Elliott's Car.* i. 45.

PLATE XVI.

Fig. 1. *Iris versicolor.*

Fig. 2. *Style and stigmas with a stamen.*

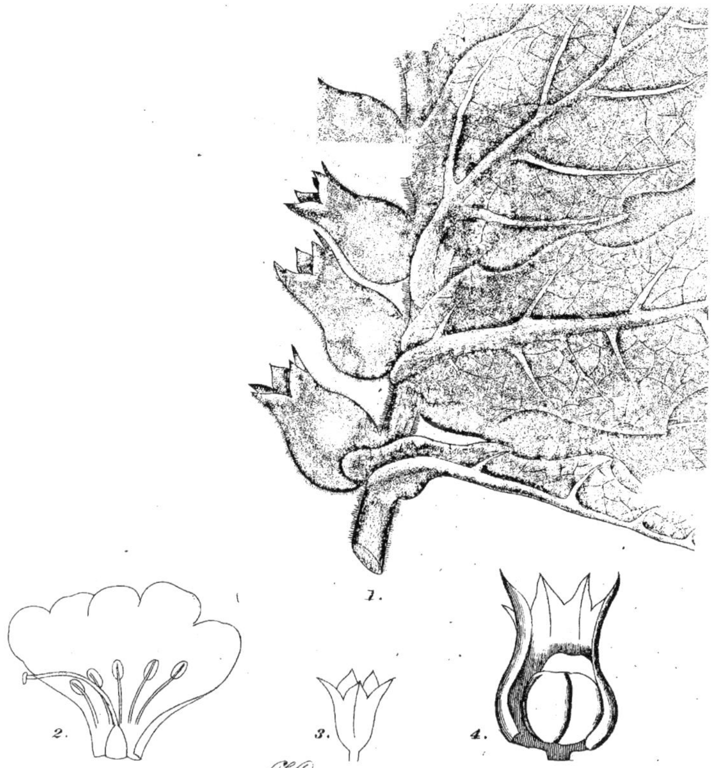

Hyoscyamus niger

HYOSCYAMUS NIGER.

Henbane.

*PLATE XVII.**

THERE is little doubt that the Hyoscyamus of this country is an imported plant. It is yet rare in most parts of the country, and appears to be strictly limited to the bounds of cultivation. Its seeds are very tenacious of life, and will spring up under favourable circumstances, after having been dormant for a long time. Hence the plant occasionally appears in old grounds which have been newly disturbed, as in grave yards, old gardens and cellars. About ten years since, a drain, which intersects the common in Boston, was opened for the purpose of repairs. In the following season a distinct row of very luxuriant plants of Henbane covered the whole of this drain, although none of them had been observed to grow in the vicinity

* For the drawing which accompanies this article, I am indebted to Dr. S. Bass of Salem.

the preceding year. The seeds, which produced these plants, had probably been buried for more than fifteen years.

This species, together with others of its genus, was well known to the ancients under the same name which it now bears. Its medicinal and deleterious properties were also understood by them.

In modern arrangements the Hyoscyamus in common with Datura, Atropa and other injurious vegetables of its kindred, is found in the class and order *Pentandria Monogynia*, and the natural order *Luridæ*. Its *Jussieuean* order is *Solaneæ*.

In this genus the *corolla is funnel shaped and obtuse, the stamens inclined, the capsule two celled and covered with a lid.*

The present species has the *lower leaves sinuated and clasping, and the flowers sessile.* It is biennial and flowers in June and July.

The whole herb has a glaucous or sea green colour, is hairy and viscid, and emits a rank, offensive smell. The stalk is one or two feet high, round, branching and rigid. The first leaves spread upon the ground, and have some resemblance to a young thistle. They are large, oblong, frequently contorted, clasping, cut into acute lobes, and pointed; the upper ones generally entire.

The flowers form a revolute, one sided spike at the end of the stem or branch, leaving, as they fall off, a straight row of capsules. The calyx has five short acute segments. The corolla is funnel shaped, irregular, with five spreading, obtuse segments, of a pale yellow or straw colour, with a beautiful net work of purple veins. Stamens inserted in the tube of the corolla, with large oblong anthers. Style slender, longer than the stamens, declined, with an obtuse stigma. Capsule two celled, roundish, covered with a lid, and invested with the persistent calyx, the segments of which extend beyond the calyx, and become rigid and prickly. The seeds are numerous, small, unequal, brownish, and are discharged by the horizontal separation of the lid.

From such chemical experiments as I have made with the dried leaves of the Hyoscyamus, I am inclined to believe, that their chief soluble portion is a variety of extractive matter. The watery and alcoholic solutions do not disturb each other, and the usual tests of tannin produce inconsiderable, or no alteration in either. Of various metallic salts which affected the solutions, the nitrate of mercury gave the largest precipitate in my experiments.

The Hyoscyamus has long been known as a narcotic poison. This character it uniformly exerts in regard to mankind, although many brute animals are exempt from its influence.* Dioscorides speaks of it as producing drowsiness and delirium.

* Horses, goats, sheep and swine are said to eat it without injury. Brute animals are frequently less susceptible of the influence of poisons than mankind. In the experiments which have been made on them to test the effect of doubtful medicines, the positive evidences of activity which they furnish, are in general more to be depended on, than the negative. That is, if an animal suffers from the action of any substance, a man would be like to suffer somewhat in the same way. Yet if the animal escapes with impunity, it does not follow that the man would be equally fortunate. There is scarcely any narcotic plant which is not devoured by some species of quadruped. Professor Pallas has stated, that the hedgehog can devour Cantharides by hundreds without inconvenience, whereas one of these insects may occasion serious trouble to a man. The following case happened under my own observation. A large eagle, (*Falco ossifragus,*) intended for a cabinet of natural history, was subjected to a variety of experiments, with a view to destroy him without injuring his plumage. A number of mineral poisons were successively given him without effect, even in large doses. At length a drachm of corrosive sublimate of mercury was inclosed in a small fish and given him to eat. After swallowing the whole of this, he continued, to appearance, perfectly well and free from inconvenience. The next day an equal quantity of arsenic was given him without any better success. So that in the end, the refractory bird was obliged to be put to death by mechanical means.

The instances recorded of deleterious consequences, ensuing from the Hyoscyamus incautiously taken, are exceedingly numerous. In a number of cases the roots have been introduced by mistake among culinary vegetables, and have occasioned alarming symptoms in whole families at once. In a case cited by Wepfer, the monks of a whole monastery, in consequence of some roots being boiled among those of chicory with their food, were seized with raving delirium, accompanied by intense thirst, impaired vision and other violent affections. Dr. Patouillat has recorded in the Philosophical Transactions, vol. 40, the case of nine persons, who were affected with loss of speech, convulsions, and at length with violent delirium. These symptoms subsided on the subsequent day, when it was found that some roots of Henbane had been dug up in the garden the preceding day by mistake for parsnips, and boiled in the soup on which the family had dined. Sir Hans Sloane, in the same Transactions for 1733, has given an instance of effects equally dangerous, occurring in some children who ate the capsules of this plant, supposing them to be filberds. Even the odour of this noxious vegetable seems capable of exciting its characteristic effects. In a case cited by Murray from the

Gazette de Sante of 1773—4, some servants who slept in a barn, where the Henbane had been scattered for a defense against rats, awoke with head-ach, dizziness and vomiting.

In instances where death has ensued from swallowing this plant, the stomach has been found to exhibit marks of inflammation, and dark or gangrenous spots.

The principal use which is made of Hyoscyamus in medicine, is as a substitute for opium, in cases where that article disagrees with the patient, or is contraindicated by particular symptoms. It appears to be free from the constipating qualities of opium, and in some instances it is found to fulfil the indications which call for an anodyne and soporific remarkably well. Among medical writers who have spoken favourably of its operation are Dr. Whytt, who employed it in various nervous diseases, and Mr. Burns, author of different obstetric works, by whom it is recommended as preferable to opium in certain puerperal complaints. Mr. Benjamin Bell states, that he found great advantage from its use in troublesome cases of chordee, after opium had failed to give relief.

It must be acknowledged however, that Hyoscyamus is far less uniform and equal in its operation than opium, and that although in some constitutions it mitigates pain, quiets irritation, and procures sleep; yet in others it produces unpleasant nausea, confusion of ideas, head-ach, and sleep which is laborious and unrefreshing. It is rather a secondary medicine, to be resorted to after the failure of opium, than one which we may confidently apply to at first, with reliance on its anodyne effects.

The Henbane was found efficacious in the disease of colica pictorum by Stoll and several others. Its external application in the form of a cataplasm of the bruised leaves has given relief in various tumors and painful affections.

For internal use the extract should be prepared in the same way as that of stramonium. From one to three grains of this extract is a suitable commencing dose, which may be gradually increased until its effects are perceived.

BOTANICAL REFERENCES.

Hyoscyamus niger, LINN. *Sp. pl.*—WOODVILLE, i. *t.* 52.—SMITH, *Engl. Bot. t.* 591.—PURSH, i. 141.—Hyoscyamus flavus, FUCHSIUS, *Hist.* 791.—Hyoscyamus vulgaris, BAUHIN, J. iii. 627.

MEDICAL REFERENCES.

STOLL, *ratio med.* iii. 13, 423.—CULLEN, *Mat. Med.* ii. 271. FOTHERGILL, *Med. Soc. Lond.* i. 310.—HOME, *clinical exp'ts,*

197.—WITHERING, *Med. Comment. Dec.* II, vi. 367.—KINGLAKE, *Lond. Med. and Phys. Journal*, v. 438.—BROWN, *ditto*, iii. 406.—MURRAY. *App. Med.* i. 655, &c. &c.

PLATE XVII.

Fig. 1. *Hyoscyamus niger.*

Fig. 2. *Corolla laid open.*

Fig. 3. *Calyx.*

Fig. 4. *Calyx of the fruit laid open to shew the capsule within it.*

1. 2. 3. 4. 5.

Solanum dulcamara

SOLANUM DULCAMARA.

Bitter sweet.

PLATE XVIII.

WHETHER the plant represented in our plate is originally of American growth, or has been introduced since the discovery of this continent, it is now difficult to say. It is certainly a hardy vegetable, and although its natural soil is at the water side, yet it easily becomes habituated to shady, fertile ground of almost any elevation. The most luxuriant specimens are found about brooks and ditches and in sheltered situations, where the roots have free access to water. In these places the stalks frequently extend some way on the surface, sending down a multitude of radicles into the mud below. When the plant grows in higher ground and more exposed to the light, its growth is restricted, and the flowers are less brilliant in colour.

The names of *Bitter sweet* and *Woody nightshade* are the most frequent English appellations of this vegetable. The former of these is also applied to the *Celastrus scandens*, a very different plant. The frequent changes which always take place in the application of vulgar names, renders a reliance on them unsafe, and indeed makes it useless to collect or preserve more than a few of the principal ones.

The genus Solanum is remarkable for the great variety and almost opposite character which takes place among its species. The common Potatoe, the Egg plant, the Tomato, the Jerusalem cherry, and the Black nightshade, are all species of this multiform genus. The common character which binds them together, consists in *a rotate corolla; the anthers cohering, with a double opening at top; the berry two celled.* The species Dulcamara is distinguished from others by *its stem, which is shrubby; unarmed and flexuous; its leaves auriculated; and its panicles resembling cymes.*

Class *Pentandria*,—Order *Monogynia*,—Natural orders *Luridæ*, Linn. *Solaneæ*, Juss.

The Bitter sweet is entitled to the character of a vine rather than shrub. The stem is woody, slender, climbing in large plants to the height of five or six feet. Leaves petioled, ovate, acute, entire, furnished at the base with two appendages, which give them somewhat of a hastate form. The lower and upper leaves are frequently without these appendages. The flowers form a loose, nodding cluster or panicle, shaped like a cyme, and taking its origin opposite to a leaf. Calyx of five short, purplish, persistent segments. Corolla rotate, becoming reflexed as it grows old, divided into five acute segments, which are purple, and marked with two whitish dots at the base of each. The filaments are much shorter than the anthers, and inserted in the short tube of the corolla. Anthers yellow, erect, cohering, so as to form a conical tube around the style. Germ oval; style longer than the stamens; stigma simple. The berries are oval, of a bright scarlet colour, and continue to hang in bunches after the leaves have fallen.

The taste and smell of the Dulcamara are less nauseous than those of many other species of Solanum. Water seems a perfect solvent for its most sensible constituents. The chief soluble portion seems to be a kind of mucous extractive, which is taken up by both water and alcohol, though most by the former. The nitrate of mercury and muriate of tin, gave precipitates from both, though most from the water. The chemical evidences of astringency were very slight.

From the experiments of Hartmann and Kuhn, cited by Murray, we may infer that water is a better solvent for this plant than alcohol. An ounce of the twigs or stalks treated with alcohol afforded two drachms and two scruples of extract. The same quantity treated with water gave three drachms and thirty five grains.

The Solanum dulcamara has formerly received the commendations of many authors, some of whose names are of high authority in medicine. The diseases in which it has acquired confidence, are chronic rheumatism, gout, secondary syphilis, incipient phthisis, asthma, jaundice. But whatever may be its efficacy in these complaints, it has in modern practice given place to more active medicines. Its most permanent and merited reputation at the present day, is derived from its application to external complaints, and particularly to cutaneous diseases. In dissertations upon the properties of this plant by Linnæus and by Carrere, its use is highly commended in herpes, in scabies, and in some of the secondary forms of syphilis. Professor Murray has added his own testimony to that of these writers, and speaks decisively of his success with it in cutaneous diseases of an inveterate character.

In the more recent and splendid works of Willan and Bateman on Diseases of the Skin, we find some important testimony of the efficacy of the Dulcamara in cutaneous affections. The former of these authors has inserted in his work a letter of Dr. Crichton, physician to the Westminster hospital, who had employed the article for a considerable number of years. This gentleman states, that out of twenty three cases of Lepra Græcorum, in which he had used it, two only had resisted its action. He does not assert that it is equally efficacious in other cutaneous diseases, although it had appeared to him to do good in psoriasis and pityriasis. His mode of employing it was as follows :

Take of stalks of Dulcamara, one ounce ; water, a pound and a half ; boil to a pound, and strain when cold.

Of this decoction the patient took two ounces at first, morning, noon and night, but the quantity was afterwards increased, until it amounted to a pint per day. At the same time the skin was ordered to be washed with a stronger decoction, which proved an auxiliary to the cure. Dr. Crichton found that in delicate people and hysterical women, it often produced syncope and slight palpitation of the heart, now and then nausea and giddiness. But if the dose was diminished, or any aromatic tincture added, it ceased to produce uneasy symptoms. The good effects of the remedy were seldom perceived until after the first eight days.

Dr. Bateman considers, that "one of the most effectual remedies for lepra under all its varieties is the decoction of the leaves and twigs of the Solanum dulcamara." He administers it in the same way with that just described. "When," says he, "there is a degree of torpor in the superficial vessels, the same decoction made with a larger proportion of the shrub, is advantageously employed as a lotion; but if there is any inflammatory disposition, this and every other external stimulus must be prohibited."

I have employed the Bitter sweet, both in substance and in decoction in a number of cutaneous affections. It appears to be a valuable auxiliary to mercury in the treatment of syphilitic eruptions. I have also known herpetic eruptions to yield to its internal and external use. The American plant however, when gathered in full vigour, does not set easily on the stomach in large doses. I have known vomiting produced by a few grains of the powdered leaves, and by a small cup of the decoction. The strength of the plant seems to vary in some degree with the time of gathering, and mode of preserving. Dr. Cullen found different parcels of the article to exhibit very different degrees of strength. Writers are not agreed as to its immediate effects on the head and stomach, probably from the different age and condition of the medicine employed by them. From my own observation I am induced to consider the appearance of slight narcotic symptoms, as an evidence of the goodness of the medicine, and as a criterion for regulating the dose. The formula of Dr. Crichton for the decoction appears to be a good one, but in the case of delicate constitutions, the commencing dose should not exceed an ounce, which may be afterwards increased according to circumstances. The addition of a little cinnamon renders the decoction less apt to offend the stomach.

BOTANICAL REFERENCES.

Solanum dulcamara, LINNÆUS, *Sp. pl.*—WOODVILLE, *t.* 33.—SMITH, *Engl. Bot. t.* 565.—PURSH, i. 156.—Solanum scandens, seu dulcamara.—TOURNEFORT, *Paris,* 43.—Glycypicros sive amaradulcis, J. BAUHIN, ii. 109.

MEDICAL REFERENCES.

Booerhaave, *Hist. hort. L. B.* 506.—Linnæus, *Amœn. Acad.* iv. 39, and viii. 62.—Murray, *App. med.* 603.—Carrere, *sur la Douce-amere,* 1780, and in *Med. and Phys. Journal,* i. 307.—Cullen, *Mat. Med.* ii. 554—Willan, *on Cutaneous diseases,* 145.—Bateman, *on ditto,* 35.—Orfila, *des poisons,* 192:

PLATE XVIII.

Fig. 1. *Solanum dulcamara.*

Fig. 2. *Corolla.*

Fig. 3. *Tube of anthers laid open.*

Fig. 4. *Stamen.*

Fig. 5. *Calyx and pistil.*

Pl. XI.

Lobelia inflata

LOBELIA INFLATA.

Indian tobacco.

PLATE XIX.

In the United States there are many species of Lobelia, which are interesting for their beauty, singularity or use. We have few plants more elegant than the cardinal flower, and few more curious in structure than the Lobelia Dortmanna. In medicinal powers, the subject of this article is entitled to take precedence of the rest. It is an annual plant, found in fields and road sides, from Canada to the southern states. It flowers from midsummer until the arrival of frosts.

The genus *Lobelia* has *a five cleft calyx; a monopetalous, irregular corolla, with a cleft tube; the anthers cohering; the capsule two or three celled.*

The species *inflata is branching and hairy, with ovate, serrate leaves, and turgid capsules.*

The connexion of the anthers into a tube has caused some ambiguity and difference of opinion, as to the place which this genus should occupy, in the Linnæan system. Linnæus placed it in his order Monogamia of the class Syngenesia. Most of our late botanists have very properly removed the plants of this order from the compound flowers, with which they have no natural affinity, to *Pentandria*, which place their number of stamens authorizes them to occupy. Pursh has placed the Lobelias under Monadelphia. The Natural order which contains them is the *Campanaceæ* of Linnæus and Jussieu.

The Lobelia inflata varies in height from six inches to two or three feet. The small plants are nearly simple, the large ones much branched. Root fibrous. Stem erect, in the full sized plant much branched, angular, very hairy. Leaves scattered, sessile, oval, serrate, veiny and hairy. Flowers in spikes or racemes, pedunculated, each one in the axil of a small leaf. Segments of the calyx linear, acute, standing on the germ, which is oval and striated. Corolla bluish purple, the tube prismatic and cleft above, the segments spreading, acute, the two upper ones lanceolate, the three lower ones oval. Anthers collected into an oblong, curved body, purple; filaments white. Style filiform; stigma curved and inclosed by the anthers. Capsules two celled, turgid, oval, compressed, ten angled, covered with the calyx. Seeds numerous, small, oblong, brown.

The Lobelia inflata when broken, emits a milky juice. When chewed, it communicates to the mouth a burning, acrimonious sensation, not unlike the taste of green tobacco. It exhibits the following noticeable ingredients upon chemical examination. 1. An acrid principle. This is evident to the taste in the tincture, decoction, and distilled water. 2. Caoutchouc. Sulphuric ether dissolves more of the plant than alcohol, and acquires a higher colour. The solution in alcohol is scarcely rendered turbid by water, that in ether is disturbed by alcohol, and grows thick as the ether evaporates. 3. Extractive. No gummy or astringent qualities were manifested in my experiments.

The great acrimony of the leaves and capsules, combined with a narcotic property, appears to be the foundation of their medicinal power. Dr. Cutler informs us, that if the leaves be held for some time in the mouth, they produce giddiness and pain in the head, with a trembling agitation of the whole body, and at length bring on nausea and vomiting. These effects are analagous to those, which the chewing or smoking of tobacco occasions in persons unaccustomed to its use.

When swallowed in substance, it excites very speedy vomiting, accompanied with distressing and long continued sickness, and even with dangerous symptoms, if the dose be large. A melancholy instance of death, occasioned by the use of this plant, in the hands of a quack, is detailed in the sixth volume of the Massachusetts Reports, in the trial of Samuel Thomson, an empiric practising in Beverly, for the murder of Ezra Lovett. In this trial it appeared, that the patient, being confined by a cold, sent for the pretended physician, who gave him three powders of Lobelia in the course of half an hour, each of which vomited him violently, and left him in a great perspiration during the night. The next day two more powders were administered, each of which operated by vomiting and occasioned great distress. In like manner two other powders were given the subsequent day, leaving the patient in a state of great prostration. Several days after this, the physician came again, and finding his patient still worse, administered several more powders, which occasioned great distress, and at length ceased to operate. Finding that the stomach was not sensible to the emetic effect of the Lobelia, the physician repeated the dose, and when the patient complained of great distress at the breast and said he was dying, the doctor assured him the medicine would soon get down, or operate as a cathartic. However, on the same evening, the patient lost his reason and became convulsed, so that two men were required to hold him. To relieve which, the doctor forced down two more of his powders, and the patient, as was to be expected, grew worse, and continued so until he expired.

The doctor, who had thus terminated the disease and the patient at once, was arrested and put upon trial for murder; but the homicide proving a legitimate one from the want of sufficient evidence of malice propense, he was acquitted and set at liberty.

From the violence of its effects, and the distressing nausea which it occasions, it is probable that the Lobelia will never come into use for the common purposes of an emetic, while other emetics can be obtained. It has however been found to exert a beneficial influence on particular diseases, and on this account is entitled to a place in the Materia Medica. Dr. Cutler, and a number of physicians in Essex county and elsewhere, have found benefit from its use in asthma, some in doses of a table spoonful of the saturated tincture, others in doses of a teaspoonful. Indeed the former dose appears to be a very large one, and greater than most stomachs would bear with impunity. I have tried this medicine in several cases of asthma with some advantage. It has not however in general succeeded in affording relief of the paroxysm, until full vomiting was produced, which effect, with me, has happened after taking one or two teaspoonfuls.

A communication from Dr. Cutler, on the operation of this plant, is inserted in Dr. Thacher's Dispensatory. The venerable writer having himself suffered from asthma for ten years, had, during the paroxysms, resorted to many medicines for relief, without experiencing much benefit from any. He was at length induced to make trial of a tincture, prepared by himself from the Lobelia inflata. "In a paroxysm," says he, "which perhaps was as severe as I had ever experienced; the difficulty of breathing extreme, and after it had continued for a considerable time, I took a table spoonful. In three or four minutes my breathing was as free as it ever was, but I felt no nausea at the stomach. In ten minutes I took another spoonful, which occasioned sickness. After ten minutes I took a third, which produced sensible effects upon the coats of the stomach, and a very little moderate puking, and a kind of prickly sensation through the whole system, even to the extremities of the fingers and toes. But all these sensations very soon subsided, and a vigour seemed to be restored to the constitution, which I had not experienced for years. I have not since had a paroxysm, and only a few times some small symptoms of asthma. Besides the violent attacks, I had scarcely passed a night without more or less of it, and often so as not to be able to lie in bed. Since that time I have enjoyed as good health, as perhaps before the first attack."

Dr. Cutler considers his disease to be what Dr. Bree in his "Practical inquiries on disordered respiration" calls the first species, "an asthma from pulmonic irritation of effused serum."

Dr. Randall informs me, that he has given the Lobelia to many persons of different ages suffering from asthma and catarrh, and with considerable variation in the form and degree of the dose. In asthma he finds it as successful as any article he has tried. When given in doses of a drachm of the saturated tincture, and two or three times repeated at convenient intervals; also in the form of other preparations of similar strength, he has found it usually to remove the paroxysm in a short time, and to restore the patient to quietude and ease. In catarrh, when given in small doses and frequently repeated, it has operated as a sure and speedy expectorant, producing effects in their most important character, very similar to those of antimony and squills. Dr. Randall has not observed any narcotic effect to ensue from moderate doses, nor found it to produce irritation of the coats of the bladder, as has been suggested by some practitioners. In his hands it has not produced any more unpleasant consequence than frequent nausea, and occasional emesis, with a copious flow from the glands of the mouth.

Dr. Bradstreet of Newburyport acquaints me, that besides asthmatic cases, he has given the saturated tincture in two or three instances of dyspepsia, also in some cases of a rheumatic nature with beneficial consequences.

He considers its sensible effects to be very like those of common tobacco, but its medicinal action more speedy and diffusible, and of shorter duration. He thinks that it affects those accustomed to the use of tobacco as readily as others.

The Lobelia has been recommended as a remedy in hooping cough and croup. In the former of these complaints, I can say nothing of its use from experience, but in the latter disease I am persuaded, it affords no benefit, having seen it largely tried by different practitioners in a number of fatal cases, where it only produced a distressing nausea, without, in any degree, facilitating the respiration, or relieving the disease.

The active properties of the Lobelia are readily extracted both by water and alcohol. The tincture however is most easily kept, and is the most convenient form for exhibition. The Essex district medical society have recommended a formula for this composition, which directs two ounces of the dried plant to be digested in a pint of diluted alcohol. Of this tincture, a teaspoonful given to an adult, will generally produce nausea, and sometimes vomiting. In certain instances however, much larger doses have been given, without producing any other effect than a flow of saliva.

BOTANICAL REFERENCES.

Lobelia inflata, LIN. *Sp. pl.*—ACT. *Upsal.* 1741, *p.* 23, *t.* 1.—GRONOVIUS, *Virg.* 134.—WILLD. *Sp. pl.* i. 946.—MICHAUX, ii. 142.—PURSH, ii. 448.

MEDICAL REFERENCES.

CUTLER, *Mem. Amer. Acad.* i. 484.—SCHŒFF, 128.—BART. *Col.* 36, 56.—THACHER, *Disp.* 267.—*Massachusetts Reports*, *vol.* vi.

PLATE XIX.

Fig. 1. *Lobelia inflata.*

Fig. 2. *Corolla with the stamens projecting from the cleft in the upper side.*

Fig. 3. *Capsule cut across.*

X

Solidago odora

SOLIDAGO ODORA.

Sweet scented Golden rod.

PLATE XX.

No part of vegetation in the United States is so conspicuous and gaudy in the autumnal months, and at the same time furnishes to the botanist so difficult a task of discrimination, as the multitudinous and Protean genera Solidago and Aster. Each of these genera contains many well defined species, sufficiently marked by their external characters, sensible qualities, habits and places of growth. But between them, is a great multitude of subspecies, liable to variation from external circumstances, changing their appearance with their places of growth, and running together by so many points of resemblance, that it is a labour yet remaining for botanists to separate those species which are in nature distinct, from those which are varieties only.

The genus Solidago is characterized by a *naked receptacle, the down simple, rays of the corolla about five, scales of the calyx imbricated and close.* It is a very natural genus, easily distinguished at sight by its crowded tufts of compound flowers, which are almost always of a deep golden yellow.*

The species *odora* has its *stem nearly smooth, leaves linear-lanceolate, entire, smooth, with a rough margin, and covered with pellucid dots. Racemes panicled, one sided.*

Class *Syngenesia*,—Order *Superflua*,—Natural orders *Compositæ*, Lin. *Corymbiferæ*, Juss.

The sweet scented Golden rod grows in woods and fields throughout the United States, and flowers in September. It has a smooth appearance, and is among the smaller species of its family. The root is woody, much branched and creeping. Stem slender, from two to three feet high, smooth or slightly pubescent below, pubescent at top. The leaves are linear-lanceolate, closely sessile, broad at base, entire, acute, with only the midrib distinct, rough at the margin but otherwise smooth, and covered with pellucid dots, like Hypericum perforatum. The flowers grow in a compound, panicled raceme, with each of its branches

* The only exception which I now recollect is *Solidago bicolor*, whose ray is white.

supported by a small leaf. These branches or peduncles are very slender and rigid, each giving off a row of ascending, downy pedicels, with small, linear bractes at their bases. Scales of the calyx oblong, acute, smooth, or slightly pubescent, the lower ones shorter and closely imbricating the rest. Florets of the ray few, with oblong, obtuse, ligules. Those of the disc funnel shaped, with acute segments. Down simple to the naked eye, feathery under the microscope. Seeds oblong.

This plant is the Solidago odora of Muhlenberg, and agrees with the character of Aiton. The Solidago odora of Michaux is possibly a different species. Willdenow's plant was undoubtedly different. The *folia puncticulosa*, which constitutes so distinct a mark in this species, I have not seen noticed by any botanist.

The leaves of the Solidago odora have a delightfully fragrant odour, partaking of that of anise and sassafras, but different from either. When subjected to distillation, a volatile oil, possessing the taste and aroma of the plant in a high degree, collects in the receiver. This oil apparently has its residence in the transparent cells, which constitute the dotting of the leaves, for the root is wholly destitute of the peculiar fragrance of the herb, and has rather a nauseous taste. This is contrary to the remark of Willdenow, who informs us that the root is the fragrant part possessing the scent of Geum urbanum.

As the volatile oil appears to possess all the medicinal value of this plant, I have not prosecuted its chemical investigation any farther.

The claims of the Solidago to stand as an article of the Materia Medica are of a humble, but not despicable kind. We import and consume many foreign drugs which possess no virtue beyond that of being aromatic, pleasant to the taste, gently stimulant, diaphoretic and carminative. All these properties the Golden rod seems fully to possess. An essence made by dissolving the essential oil in proof spirit, is used in the eastern states as a remedy in complaints, arising from flatulence, and as a vehicle for unpleasant medicines of various kinds. I have employed it to allay vomiting, and to relieve spasmodic pains in the stomach of the milder kind, with satisfactory success. From its pleasant flavour, it serves to cover the taste of laudanum, castor oil, and other medicines, whose disagreeable taste causes them to be rejected by delicate and irritable stomachs.

Mr. Pursh informs us, that this plant when dried, is used in some parts of the United States as an agreeable substitute for tea. He further states, that it has for some time been an article of exportation to China, where it fetches a high price.

BOTANICAL REFERENCES.

Solidago odora, AITON, *Hort. Kew.* iii. 214.—PURSH, ii. 539.—Virga Aurea Americana, Tarraconis facie et sapore, panicula speciosissima? PLUKENET, *Alm.* 389, *t.* 116, *f.* 6.

PLATE XX.

Fig. 1. *Solida odora.*

Fig. 2. *A flower magnified.*

Fig. 3. *A floret of the ray.*

Fig. 4. *A floret of the disc.*

NOTES.

Note A.

Most European writers seem to consider the Datura stramonium as a native of America. In Miller's Dictionary by Martyn, the editor says, "That it is a native of America, we have the most undoubted proofs, for in earth brought with plants from various parts of that extensive country, we are sure to have the Thorn apple come up. Kalm says, that it grows about all the villages, and that this and the Phytolacca are the worst weeds there. Our old writers call it Thorny Apples of Peru."

This evidence however is by no means sufficient. The plant appears in earth and ballast, carried from either continent alike. The name Apple of Peru has also been applied to Datura metel, a plant of Africa and the East Indies.

Note B.

In the Catalogue of plants in the Botanic garden at Calcutta, published in 1814, a species is inserted by the name of Datura Tatula, said to be a native of the Cape of Good Hope. This is probably different from the Datura Tatula of Linnæus.

Note C.

"The Jamestown weed, (which resembles the thorny apples of Peru, and I take it to be the plant so called,) is supposed to be one of the greatest coolers in the world. This being an early plant, was gathered very young for a boiled sallad, by some of the soldiers sent thither to quell the rebellion of Bacon; and some of them ate plentifully of it, the effect of which was a very pleasant comedy, for they turned natural fools upon it for several days. One would blow up a feather in the air, another would dart straws at it with much fury; another stark naked was sitting up in a corner like a monkey, grinning, and making mows at them; a fourth would fondly kiss and paw his companions, and sneer in their faces with a countenance more antic, than any in a Dutch droll. In this frantic condition they were confined, lest, in their folly, they should destroy themselves. A thousand simple tricks they played, and after eleven days returned to themselves again, not remembering any thing that had passed."

Beverly's History of Virginia, p. 121.

Note D.

" De Cuechyliztomatl, seu Tomatl sonalis.

Genus est Solani Tonchichi forma et viribus simile, sed foliis paulisper undulatis, et fructu acinoso racematimque dependente, &c." *Hernandez,* ii. 12.

Note E.

" I am heartily glad to hear more instances of the success of the Poke weed in the cure of cancer. You will deserve highly of mankind for the communication. But I find in Boston they are at a loss to know the right plant, some asserting it is what they call Mechoacan, others other things. In one of their late papers it is publicly requested that a perfect decription may be given of the plant, its places of growth, &c. I have mislaid the paper, or would send it to you. I thought you had described it pretty fully." *Letter from Dr. Franklin to Dr. Colden.*

" I apprehend that our poke-weed is what botanists term phytolacca. This plant bears berries as large as peas. The skin is black, but it contains a crimson juice. It is this juice thickened by evaporation in the sun which was employed. It caused great pain, but some persons were said to have been cured. I am not quite certain of the facts; all that I know is that Dr. Colden had a good opinion of the remedy." *Letter from Dr. Franklin to M. Dubourg.*

Note F.

Linnæus, in his Flora Laponica, tells us that the roots of *Calla palustris*, although acrid and caustic in the highest degree, (*ignis firme instar,*) are made into a kind of bread in high estimation, called *Missebrœd.* This is performed by drying and grinding the roots, afterwards boiling and macerating them until they are deprived of acrimony, when they are baked like other farinaceous substances into bread.

The recent juice of the *Jatropha manihot*, or Cassava tree of the West Indies, is highly poisonous. The deleterious principle however resides in a volatile portion, which is dissipated by heat. The remaining substance of the root is used by the inhabitants for bread, as a material for a kind of soup, and as the basis of a fermented liquor.

Note G.

The following is Kæmpfer's description taken from his Amœnitates Exoticæ, p. 791. His accompanying figure resembles the American Rhus vernix, except, that the end of the branch and bud are larger in proportion than with us.

" *Sitz*, vel. *Sitzdsju*, i. e. *Sitz* planta, vulgo *Urus* seu *Urus no ki*, Arbor vernicifera legitima, folio pinnato Juglandis, fructu racemoso ciceris facie.

" Arbor paucis ramis brachiata, salicis ad altitudinem luxuriose exsurgit. Cortice donatur incano, ex verruculis scabro, facile abscedente; *ligno* saligneo fragillimo; *medulla* copiosa, ligno adnata; *Surculis* longis crassis in extremitate inordinate foliosis. *Folium* est impariter pennatum, spithamale vel longius, Juglandis folio æmulum, costa tereti, leviter lanuginosa; quam a semipalmari nuditate stipant lobi sive folia simplicia, pediculo perbrevi nixa, tenuia, plana, ovata, trium vel quatuor unciarum longitudinis, basi inæqualiter rotunda, mucrone brevi angusto, margine integro, suprema facie obscure viridi, lævi, et ex nervis lacunosa, dorso incano et molliter lanuginoso. Nervus medius in mucronem terminans subinde multos a latere demittit nervos minores, citra marginem deficientes. Sapor folio sylvestris inest, cum sensibili calore; humor affrictus extemplo chartam ferrugineo colore imbuit. In surculis quibusdam ex foliorum axillis singuli surgunt *Racemi* laxe ramosi, palmares, tenues, qui, petiolis in calyculos rotundos desinentibus, *Flosculos* continent pumilos, et citra Coriandri seminis magnitudinem radiantes, in luteum herbaceos, pentapetalos, petalis carnosis nonnihil oblongis et repandis; staminibus ad petalorum interstitia singulis, apicatis, brevissimis, stylo perbrevi tricipite, floris turbini insidente. Ordorem spirant dulcem, Aurantio flori affinem et pergratum. *Fructus* flosculum excipit gibbosus, utcumque in rhomboidis figuram compressus, bifidus, facie ac magnitudine ciceris, membranula tenui micante vestitus, per maturitatem durissimus et obsoleti coloris.

"Cortex arboris cultro crenatus lacteum fundit lentorem, humore crystallino (ex aliis ductibus stillante) permixtum, qui ad aeris contactum nigrescit. Eundem surculi divulsi, foliorum pediculi, et nervi produnt, nullius gustabilis qualitatis participem, nisi califacientis sine acredine. Venenatos tamen spiritus hæc arbor exhalare dicitur, vehementes adeo, ut pueris circa eandem commorantibus exanthemata in corpore pariant: qualia etiam lignum tractantes alii (non omnes) experiuntur. Collectio *Urusj*, sive Vernicis, ut instituatur, caudices præcipue triennes, paucis crenis vulnerandæ sunt, ex quibus stillans liquor subinde excipitur, iterata in recente loco sectione, donec exsucci marcescant. Emulsi atque omni succo orbati, illico amputandi sunt; sic nova e radice provenit soboles, quæ, triennis facta, collectioni denuo subjicitur." * * *

"Vernix nativa vix præparatione indiget. Japonica per duplicatam chartam subtilissimam, telæ aranearum pene similem, et eam in rem singulariter constructam docta *εγχέιρησι* torqueri solet, ut a particulis heterogeneis et crassioribus mundetur; mundatæ pauxillum admiscetur (centissima fere pars) olei *Toi* dicti ex fructu arboris *Kiri*. Sic vasibus ligneis indita per Japoniam venalis transvehitur."

Note H.

The following account of the death of Socrates is translated from the *Phœdon* of Plato.

And Crito hearing this gave the sign to the boy who stood near. And the boy departing after some time returned bringing with him the man, who was to administer the poison, who brought it ready bruised in a cup. And Socrates beholding the man, said, "Good friend, come hither, you are experienced in these affairs,—What is to be done?" "Nothing," replied the man, "only when you have drank the poison, you are to walk about until a heaviness takes place in your legs. Then lie down. This is all you have to do." At the same time he presented him the cup. Socrates received it from him with great calmness, without fear or change of countenance, and regarding the man with his usual stern aspect, he asked, "What say you of this potion? Is it lawful to sprinkle any portion of it on the earth as a libation, or not?" "We only bruise," said the man, "as much as is barely sufficient for the purpose." "I understand you," said Socrates, "but it is certainly lawful and proper to pray the gods that my departure from hence may be prosperous and happy, which I indeed beseech them to grant." So saying, he carried the cup to his mouth and drank it with great promptness and facility.

Thus far most of us had been able to refrain from weeping. But when we saw that he was drinking and actually had drunk the poison, we could no longer restrain our tears. And from me they broke forth with such violence, that I covered my face and deplored my wretchedness. I did not weep for his fate, so much, as for the loss of a friend and benefactor, which I was about to sustain. But Crito unable to restrain his tears was compelled to rise. And Apollodorus, who had been incessantly weeping, now broke forth into loud lamentations, which infected all who were present except Socrates. But, he observing us, exclaimed, "What is it you do, my excellent friends? I have sent away the women that they might not betray such weakness. I have heard that it is our duty to die cheerfully and with expressions of joy and praise. Be silent therefore, and let your fortitude be seen." At this address we blushed and suppressed our tears. But Socrates, after walking about, now told us that his legs were beginning to grow heavy, and immediately laid down, for so he had been ordered. At the same time the man who had given him the poison, examined his feet and legs, touching them at intervals. At length he pressed violently upon his foot, and asked if he felt it. To which Socrates replied, that he did not. The man then pressed his legs and so on, shewing us that he was becoming cold and stiff. And Socrates feeling of himself assured us, that when the effects had ascended to his heart he should then be gone. And now the middle of his body growing cold, he threw aside his clothes and spoke for the last time, "Crito, we owe the sacrifice of a cock to Æsculapius. Discharge this and neglect it not." "It shall be done, said Crito; have you any thing else to say?" He made no reply, but a moment after moved, and his eyes became fixed. And Crito seeing this, closed his eyelids and mouth.

CONTENTS OF THE FIRST VOLUME.

ADVERTISEMENT.

UPON presenting to the public the portion of this work, which completes one half of his labours, the author of the American Medical Botany feels himself bound to offer some report of the progress and prospects of his undertaking.

The plan of this work was formed and announced at a time when such a subject was wholly novel, and when coloured botanical engravings, from the difficulty and expense of their execution, were almost unknown in this country. It was endeavoured that the plan should be such as in some degree to insure, what both authors and their friends should desire, extensive circulation and permanent utility. An edition of a thousand copies was printed, and that the work might be generally accessible, the price was fixed at a lower rate, in proportion to the number of plates, than works of the kind have ever been placed at in this country; a rate which the probable sale of the whole edition could alone justify. As permanent utility was esteemed of more importance than ephemeral success, the work has not been hurried in any stage of its execution, even though some temporary advantages might have been obtained by its earlier appearance. The author has not willingly adopted the opinions of others in botany or medicine without examination, and has thought no delay injurious which might lead to the establishment of truth or the detection of error. He has been desirous, in adding the results of his own experience, that the book should have, in some degree at least, the character of an original work, rather than of a compilation; at the same time that it should present a fair view of what is known on the subjects of which it treats. The figures of the plants have, in every instance, been made from original drawings, which were executed by himself, with the exception of two or three presented by his friends. The style of engraving is wholly new in this country, and is one which has been successfully attempted only by the first artists in France.

It gives him pleasure to state, that the reception of the work, in all parts of the United States, has exceeded his anticipations, that the subscription is already more than sufficient to defray the expense of publishing, and that its regular increase renders it probable that the whole edition will be taken up at an early period.

He avails himself of this occasion to return his acknowledgments to those correspondents who have obligingly assisted him by the communication of specimens for the work, and of the results of their own researches and experience. Particularly he would express this remembrance to Zaccheus Collins Esq. of Philadelphia, a gentleman whose active kindness has repeatedly supplied his botanical necessities; and whose extensive erudition has enabled him to afford counsel, which would not have been sought at a less respectable source. To Professor Ives of New Haven, a zealous and intelligent cultivator of the American Materia Medica, he would express the obligations derived from his communications and correspondence. It would be unjust to forget that many medicinal plants of the Southern States, with observations on their properties, were furnished expressly for this work, by the late Dr. James Macbride of Charleston, S. C. a physician and a botanist, whose premature death has terminated a career of honourable usefulness and of active, liberal and efficient prosecution of science. His friends cannot remember without regret a man, who had the rare quality of being learned without ostentation, who was ambitious of usefulness more than of fame, and who sought rather to be valuable to others than just to himself.

As the materials for the rest of the American Medical Botany are now principally collected, and most of the drawings finished; the remaining numbers will be issued with as much promptness and regularity, as is consistent with their faithful execution.

PREFACE

TO THE SECOND VOLUME.

As frequent use is made in these pages of observations drawn from the auxiliary sciences, as affording some light on the medicinal properties of plants, it may be proper to examine how far testimony of this kind is entitled to receive credit in our inquiries and examinations.

There can be no question, that the actual operation of medicines upon the human system, gathered from positive experience, is, in the present state of our knowledge, the only criterion by which we can pronounce, with universal certainty, on their properties. There are nevertheless many things to be learnt from chemical analysis, sensible qualities, and botanical affinity, which may afford us, in some instances certainty, and in most others presumptive evidence of the medicinal characters of vegetables. The correspondence in these respects is frequently so striking, that we can hardly resist the belief, that an entire harmony of properties exists, which, if we are unable

2

fully to comprehend, it is rather owing to the imperfection of science, than to the irregularity of nature.

A few illustrations of this point, taken from general facts already ascertained, will place the subject in a clearer light.

The chemical substances, known by the names of *Gum Mucus* and *Fæcula*, are constantly emollient, demulcent, and nutritious. They manifest these qualities even when extracted from acrid and poisonous vegetables, as in Arum, Calla, and Jatropha.

Sugar is nutritious and demulcent. When subjected to a spontaneous chemical change by the vinous fermentation, it is universally a strong diffusible stimulus.

Fixed oils are emollient and laxative. Also nutritious.

Volatile oils on the contrary are acrid, stimulating, heating, and antispasmodic.

Tannin and the *Gallic* acid are uniformly antiseptic and powerfully astringent.

The *Acetous*, *Citric*, *Tartaric* and similar vegetable *acids* are refrigerant and antiseptic.

Bitter *Extractive* substances are usually tonic.

Resins, which are bitter and acrid, are commonly cathartic.

Emetine, as separated by Pelletier and Magendie, is powerfully emetic.

Morphium, obtained by Serturner, is a very strong narcotic.

The foregoing are some of the examples, which the present state of Chemistry allows us to observe of affinity between chemical and medicinal characters. With a few exceptions they will be found to be strictly true. Yet the analysis of vegetables is at present but imperfectly known, and an extended investigation is continually bringing new principles to light. We can hardly expect that the business of generalization should be attempted with complete success, until the constituent facts are better understood. From what we already know, however, it is not chimerical to predict, that if the chemistry of vegetables were as perfectly known in all its parts, as in those which we have detailed; their medicinal properties might be inferred, with at least as great certainty, as that which now attends most inferences in the conjectural science of medicine.

In regard to the botanical affinities of plants, as affording evidence of their medicinal powers, much has been said and written. Petiver, Hoffman, Linnæus, Hasselquist, and recently the learned Professor Decandolle have bestowed much investigation on this subject. It is regarded as a desideratum by all, and as the consummation of botanical science by many, that plants should be so arranged, as that their assemblages should agree, not only in external forms, but in internal qualities and operative powers. Certain general agreements of this kind evidently prevail throughout nature; yet they are so varied, and subject to so many exceptions, that it is difficult to establish them by general scientific descriptions, and when they are rendered too minute they seem to lose much of their importance. It is perhaps as easy to know the properties of plants from their external habit, as to understand the characters of mankind from their physiognomy. Accurate observers know more than they can communicate the means of knowing to others, yet the most accurate are liable to be mistaken. Many vegetables of the closest affinity and resemblance, even species of the same genus, differ wholly from each other in their effects. Witness the species of Cucumis, Convolvulus, and Solanum, some of which are salutary, and others highly deleterious. Nevertheless there are many general truths, or at least general probabilities, by which every one would be influenced, and which have so much importance, that they will never be forgotten.

No botanist, even if in danger of starving in a wilderness, would indulge his hunger on a root or fruit taken from an unknown plant of the natural order *Luridæ*, of the *Multisiliquæ*, or the *umbelliferous aquatics*. On the contrary, he would not feel a moment's hesitation in regard to any of the *Gramina*, the fruit of the *Pomaceæ*, and several other natural families of plants, which are known to be uniformly innocent in their effects.

The *sensible properties* of plants afford another clue to their influence on the human system. It is true, that observations derived from this source will not serve us in forming very minute distinctions. They are, however, almost always adequate in vegetable productions, to enable us to distinguish what is innocent and salubrious, from what is noxious and virulent. The brute creation depend wholly upon the powers of sense in selecting their food, and this reliance does not often betray them. In regard to mankind it almost uniformly happens, that what is sweet, delicious, or aromatic, proves nutritive or salutary; while on the other hand, vegetable poisons are nauseous, acrid, and disgusting. It has been observed, that it would have been a sort of treachery in nature to have made it otherwise. Considering the universal dissemination of poisonous plants, and the number of them, which frequent the vicinity of human habitations, this arrangement of Providence, by making ungrateful what is dangerous, has furnished almost the only safeguard from harm, to the inexperienced and unwary.

These remarks have been offered on account of an impression which many persons entertain, that collateral evidences of the characters of plants are worthless and undeserving of attention. Even if the community were composed exclusively of physicians, such an opinion could not be wholly correct. Every one may be called on to form hasty decisions on subjects where his experience is deficient, and where an acquaintance with auxiliary facts might lead him to a correct issue. It is not only curious and instructive to perceive the harmonies of nature, but to every inquirer among her works it must be practically useful. It can no where be more useful, or more deserving of study, than in a new country, where the face of nature presents an ungathered harvest, and where every clue to useful discovery derives importance from its influence and tendency.

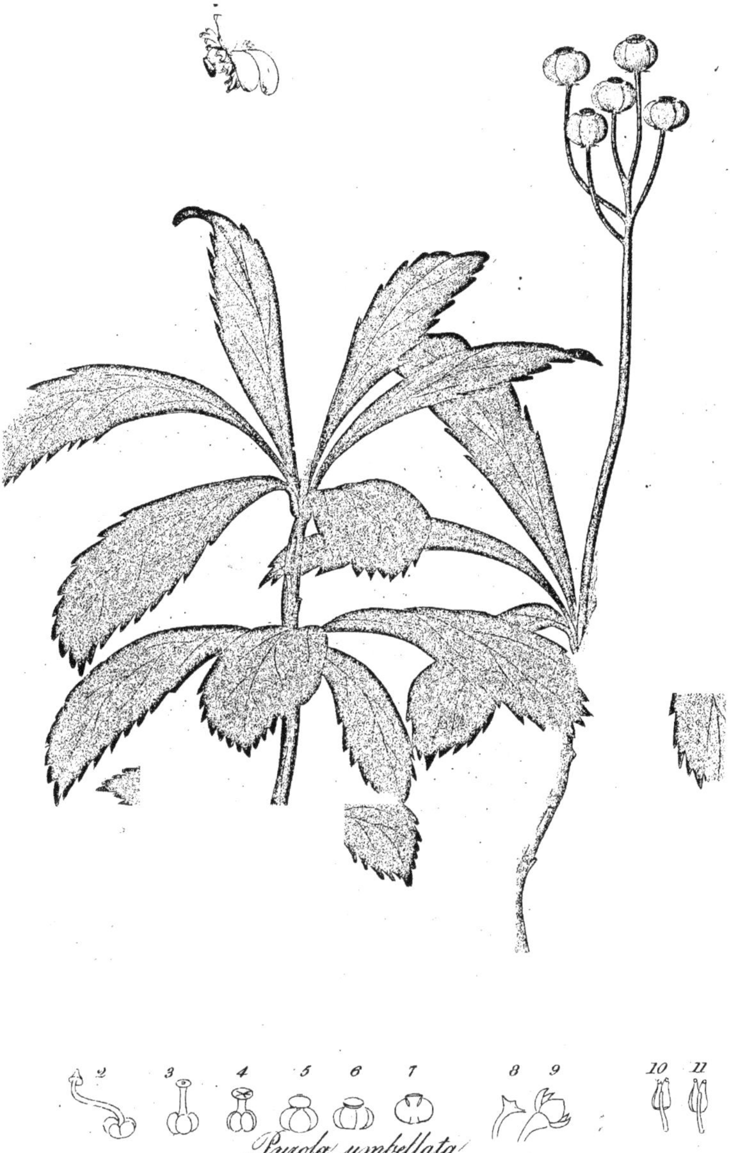

Pyrola umbellata

AMERICAN

MEDICAL BOTANY.

PYROLA UMBELLATA.

Winter Green.

PLATE XXI.

THIS most beautiful of the species of Pyrola is extensively diffused throughout the northern hemisphere. It inhabits all latitudes in the United States, and extends across the continent to the shores of the Pacific ocean. It is also found in the forests of Siberia, and in several of the northern and temperate countries of Europe. It only grows in shady woods, where it is protected from the sun, and nourished by the peculiar soil formed from the decomposition of leaves and wood. The most common appellations, by which it is known in the United States, are *Winter green* and *Pipsissewa.* It flowers in June and July, being somewhat later than most of the other species of its family.

By Pursh and some other American botanists, this species and one other have been separated from the genus Pyrola, to constitute a new family by the name of Chimaphila. As the grounds of distinction, however, between them are not sufficient to render it certain that this genus will ultimately stand; I have preferred retaining the original Linnæan name.*

* It is somewhat remarkable, that the genus Chimaphila was first established upon characters, which hardly exist in either of the plants it is intended to comprehend. The principal grounds of distinction, suggested by Michaux and adopted by Pursh, seem to consist in a *sessile stigma*, and anthers opening by a *subbivalve foramen.* Now the stigma is not sessile, since that term implies the absence of a style, and the anthers do not open by any subbivalve foramen, differing from the rest, but by two tubular pores, precisely as in the other species of Pyrola. Mr. Nuttall, in his interesting work on North American genera, has amended the character of Chimaphila, by bringing into view the calyx, filaments, &c. while he has added to the characteristics of Pyrola, a downy connexion of the valves of the capsule. In the calyx, however, the two species of Chimaphila are at different extremes from each other; one of them having a five leaved calyx, the leaves overlaying each other at base; the other having a five toothed calyx only, while the remaining species of Pyrola, being five parted, come between them. I have not been able to find the *tomentum* spoken of by Mr. Nuttall, in all the spiked species, and particularly in P. secunda.

If the genus Pyrola were ever to be dismembered, it should be into at least four distinct genera, as follows;

1. *Style declined, stigma annulate.*

P. rotundifolia, P. asarifolia, &c.

The genus Pyrola belongs to the class *Decandria*, and order *Monogynia*. It ranks among

2. *Style straight, stigma peltate.*
   P. secunda, P. uniflora, &c.
3. *Style incrassated, calyx five leaved.*
   P. maculata.
4. *Style immersed, calyx five toothed.*
   P. umbellata.

If we go farther and take into view the direction and form of the filaments, and the other parts of flower and fruit, with their various combinations; we shall have nearly as many genera as there are now species, since it is well known that many of the most important specific distinctions in this genus are taken from the fructification.

On these accounts there can be no doubt that the genus Pyrola had better remain entire. In habit it is certainly one of the most natural genera we possess. All the species are humble evergreens, growing in woods, with creeping roots, ascending stems, and nodding flowers. All of them have their leaves in irregular whorls, flower with reversed anthers, and retain their style until the fruit is ripe. In inflorescence, one is solitary, two somewhat corymbed, and the rest spiked. The leaves of P. secunda, umbellata and maculata are usually in two or more whorls; those of most others in one radical whorl or aggregate. One species is said to be leafless.

In the dissections accompanying the figure of P. umbellata I have endeavoured to represent the evident gradation of the style from the species in which it is longest, to that in which it is shortest. In the same plate are added some of the varieties of the calyx and stamens.

The following remark of Sir James Edward Smith, the learned president of the Linnæan society, is from Rees' Cyclopedia, *Art.* PYROLA. "We can by no means assent to the establishment of that able writer's *(Fursh's)* Genus Chimaphila, there being surely no diversity of habit to support it, nor any character but a difference in the length of the style; which the other species of Pyrola shew to afford admirable specific, but no generic distinctions.

the *Bicórnes* of Linnæus and the *Ericæ* of Jussieu.

The generic character is as follows. *Calyx mostly five parted; petals five; anthers inverted, opening by two tubular pores; capsule five celled, five valved.*

The species *umbellata* has its *leaves wedge shaped and toothed, flowers somewhat umbelled, calyx five toothed, and style immersed.*

Its more minute description is as follows:

Root woody, creeping, sending up stems at various distances. The stems are ascending, somewhat angular, and marked with the scars of the former leaves The leaves grow in irregular whorls, of which there are from one to four. They are evergreen, coriaceous, on very short petioles, wedge shaped, subacute, serrate, smooth, shining, the lower surface somewhat paler. The flowers grow in a small corymb, on nodding peduncles, which are furnished with linear bractes about their middle. Calyx of five roundish subacute teeth or segments, much shorter than the corolla. Petals five, roundish, concave, spreading, cream coloured, with a tinge of purple at base. Stamens ten. Filaments sigmoid, the lower half fleshy, triangular, dilated, and slightly pubescent at the edges; the upper half filiform. Anthers two celled, each cell opening by a short, round, tubular orifice, which points downward in the bud, but upward in the flower. Pollen white. Germ roundish, depressed, furrowed, obscurely five lobed, with a funnel shaped cavity at top. Style straight, half as long as the germ, inversely conical, inserted in the cavity of the germ, and concealed by the stigma. Stigma large peltate, convex, moist, obscurely five rayed. Capsules erect, depressed, five celled, five valved, the partitions from the middle of the valves. Seeds linear, chaffy, very numerous and minute.

This plant, like the other species of Pyrola, is very difficult to cultivate, when transplanted from its native soil; although it thrives luxuriantly in the shade and rich mould of the forests where it originates.

The leaves of Pyrola umbellata, when chewed, communicate to the mouth a taste which partakes of both sweet and bitter. The stalk and roots possess the same taste, combined with a moderate degree of pungency. A Dissertation "De Pyrola umbellata," published at Gottingen, by Dr. Wolf, in 1817, contains an elaborate chemical examination of this plant. As the result of his trials, this author concludes, that 100 parts of Pyrola umbellata contain about 18 of a bitter extractive principle, 2.04 of resin, 1.38 of tannin, a slight portion of gum, and the rest of fibrina and earthy salts. The resin is adhesive, brownish, readily soluble in ether and alkalis, burning with flame and a resinous odour, and leaving a white cinder.

From my own trials the quantity of resin in this plant appears to be very small. A saturated tincture of a deep brown colour does not give a precipitate on the first addition of water. It is only after some time standing, and partly perhaps from the evaporation of the alcohol, that a turbidness begins to appear in the solution. It is probable that spirit is a better menstruum than water for the soluble portions of this plant, although the latter is capable of extracting the greater part of its virtue.

The Pyrola umbellata, though scarcely known as a medicine until within a few years past, has at the present day, acquired a reputation of considerable extent in the treatment of various diseases. Its popular celebrity seems to have originated in its application to the treatment of fever and rheumatism; but the attention of physicians has been chiefly drawn towards its use in other complaints. The instances in which this plant has received favourable testimonies on medical authority, of its successful use, both in America and Europe, are principally the following. 1. As a palliative in strangury and nephritis. 2. As a diuretic in dropsy. 3. As an external stimulant, susceptible of useful application to various cases.

In the first of these cases, the Pyrola is entitled to attention and confidence. Some practitioners in this country have employed it with advantage in the same cases, in which the Arbutus Uva ursi is recommended*. Dr. Wolf, the German writer lately cited, has reported a number of cases of ischuria and dysuria, arising from various causes, in which the Pyrola, given in infusion, produced the most evident relief, and took precedence of a variety of remedies which had been tried. His method of administering it was to give a table spoonful of a strong infusion, with a little syrup, every hour. In all the cases he has detailed, small as the dose was, it gave relief in a very short time. In one case its effect was so distinctly marked, that the disease returned whenever the medicine was omitted and was removed on resuming its use. A tonic operation attended its other effects, so that the appetite was improved and digestion promoted during the period of its employment.

* See Dr. Mitchell's Inaugural Dissertation. Philadelphia, 1803.

The diuretic properties of the Pyrola umbellata, seem to have been fully illustrated by Dr. W. Somerville in a paper on this vegetable, published in the 5th volume of the London Medico-Chirurgical transactions. The facts presented by this physician afford satisfactory evidence of the power of this medicine to promote the renal excretion, and to afford relief to patients afflicted with dropsy in its various forms. The most distinguished case presented by him, is that of Sir James Craig, the British governour in Canada, who was labouring under a general dropsy, which in its progress had assumed the forms of hydrothorax, anasarca and ascites, and which was combined with different organic diseases, especially of the liver. After having tried with little or temporary success, almost every variety of diuretic and cathartic medicines, and submitted twice to the operation of tapping, the patient had recourse to a strong infusion of the Pyrola, in the quantity of a pint every twenty four hours. Although the case was altogether an unpromising one, yet the plant gave relief, not only in the first, but in the subsequent instances of its use. It increased the urinal discharge, and at the same time produced an augmentation of strength and an invigorated appetite.

Several other cases of dropsy are detailed in Dr. Somerville's paper, in which the Pyrola was administered by himself and by other practioners with decided advantage. Dr. Satterly and Dr. Marcet are among those who have added their observations to the testimonies in its favour. Dr. Somerville found his patients to remark, that an agreeable sensation was perceived in the stomach soon after taking the Pyrola, and that this was followed in some instances by an extraordinary increase of appetite. He considers it as having in this respect a great advantage over other diuretics, none of which are agreeable to the stomach, and most of them very offensive to it. He further states, that no circumstance had occurred within his own experience or information, to forbid its use in any form, or to limit the dose.

Dr. Wolf has given one very satisfactory case of the utility of our plant in ascites. He also found it to alleviate altogether the ardor urinæ attendant on gonorrhea.

Such are the most important facts which to my knowledge have been published respecting the internal use of the Pyrola umbellata. I have administered this plant on various occasions, and attended to its mode of operation. In a number of dropsical cases, when first given, it made a distinct and evident impression on the disease, communicating an increased activity to the absorbents, followed by a great augmentation of the excretion from the kidnies. The benefit, however, with me has been in most instances temporary, and it was found better to omit the medicine for a time and to resume it afresh, than to continue it until the system had become insensible to its stimulus. After suspending it for a week or two, the same distinct operation took place on returning to its use, as had been manifested in the first instance. It proved in almost every instance, a very acceptable medicine to the patient, and was preferred both for its sensible qualities and its effects on the stomach, to other diuretics and alteratives which had been prescribed.

The Pyrola has been considerably employed as an external application in tumours and ulcers of various descriptions. It first acquired notice in consequence of some newspaper attestations of its efficacy in the cure of cancer. Those persons who know how seldom genuine cancers occur in comparison with reputed ones, will be more ready to allow it the character of curing ulcerous, than really cancerous affections. There are undoubtedly many ulcers, and those frequently of a malignant kind, which are benefitted by antiseptic stimulants ; and to such the Pyrola may be useful. But of its efficacy in real cancer we require more evidence than is at present possessed, before we ascribe to it the power of controlling so formidable a malady.

Dr. Miller of Franklin informs me that he has used a decoction and cataplasm of this plant with apparent success in various chronic indurated swellings. It acts as a topical stimulant, and when long continued, not unfrequently vesicates. Tumours of long standing have in several instances disappeared under its use.

BOTANICAL REFERENCES.

Pyrola umbellata, LIN. *Sp. pl.* GMELIN, *Flora Sibirica.* ROTH, *Flora Germanica.*—*Bot. Mag. t.* 778.—MICHAUX, *Flora Americana,* i. 251.—Pyrola fruticans, PARKINSON, *Theatrum,* 509.—J. BAUHIN, *Hist. plant.* iii. 536.—Chimaphila corymbosa, PURSH, i. 300.—NUTTALL, *Genera,* i. 274.

MEDICAL REFERENCES.

MITCHELL, *Inaugural Dissertation.*—SOMERVILLE, *Medico-Chirurgical Transactions,* vol. v.—WOLF, *Dissertatio Inauguralis.*

PLATE XXI.

Fig. 1. *Pyrola umbellata.*
Fig. 2. *Pistil of Pyrola rotundifolia.*
Fig. 3. *Pistil of Pyrola secunda.*
Fig. 4. *Pistil of Pyrola uniflora.*
Fig. 5. *Pistil of Pyrola maculata.*

Fig. 6. *Pistil of Pyrola umbellata.*

Fig. 7. *Section of the same, shewing the length of the style.*

Fig. 8. *Five toothed calyx of* **P.** *umbellata.*

Fig. 9. *Five leaved calyx and incrassated pedicel of* **P.** *maculata.*

Fig. 10. *Anther magnified of* **P.** *secunda.*

Fig. 11. *Ditto of* **P.** *rotundifolia.*

Fig. 12. *Stamen magnified of* **P.** *umbellata.*

Fig. 13. *Ditto of* **P.** *maculata.*

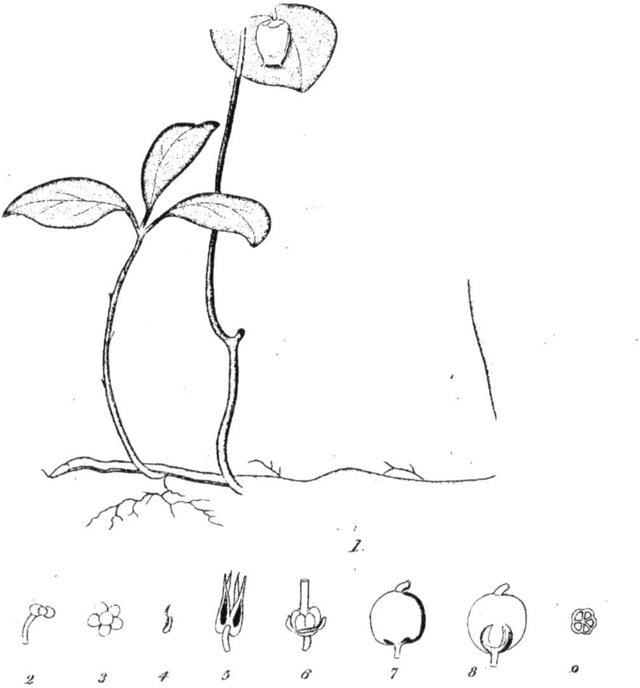

Gaultheria procumbens.

GAULTHERIA PROCUMBENS.

Partridge Berry.

PLATE XXII.

There is no soil so inhospitable, that it does not afford the means of sustenance and growth to some vegetable tenant. The most arid and penurious spots of earth not only give support to a variety of plants, but they are even selected by certain species, which make them their permanent residence, and thrive better in the midst of poverty and drought than they could in the most fertile and luxuriant situations. The Gaultheria procumbens is one of those hardy and abstemious plants, which are better satisfied with the clear air of the mountains, than with a deep or mellow soil. It is found growing in large beds under the shade of shrubs and trees upon elevated tracts of ground, or upon the sand and gravel of the driest forests. Its bright evergreen leaves seem adapted for ready absorption and slow perspiration, so that it derives from the dews and rain, what the earth fails to supply it.

The Gaultheria procumbens is remarkable for the different periods of producing its flowers and fruit. It is found in blossom not only in the early part of spring, but in the last weeks of summer, and the fruit is found ripe at corresponding periods. Whether this appearance is the product of different shoots, or whether the same stems blossom twice in a year, I am unable to say. I have, however, met with beds of the Gaultheria in full flower in August and September, quite as frequently as in May. I have also seen the fruit in the market at various periods of the summer, fall, and spring.

The plant takes its vulgar names from the fruit, and is denominated in different parts of the United States, *Partridge berry*, *Chequer berry*, *Box berry*, &c. Its domestic use has also given it the name of *Mountain tea.*

The genus Gaultheria is beautifully singular and distinct in its character, derived from the form of its fruit. The *calyx is five cleft, calyculated, or bibracteate at base. Corolla ovate. Capsule five celled, invested with the baccated calyx.*

The species *procumbens* has a *prostrate stem with ascending branches. Leaves in a terminal tuft, obovate with a few ciliate serratures. Flowers axillary.*

Class *Decandria*, order *Monogynia*, Natural orders *Bicornes Linn. Ericæ Juss.*

The stem, or as it might be called root of this plant is horizontal, woody, often a quarter of an inch in thickness. The branches are ascending, but a few inches high, round and somewhat downy. Leaves scattered, near the extremities of the branches, evergreen, coriaceous, shining, oval or obovate, acute at both ends, revolute at the edge, and furnished with a few small serratures, each terminating in a bristle. Flowers axillary, drooping, on round downy stalks. Outer calyx of two concave, heart shaped leafets, which may with perhaps more propriety be called bractes. Inner calyx monophyllous, white, cleft into five roundish subacute segments. Corolla white, urceolate, five angled, contracted at the mouth, the border divided into five short, reflexed segments, Filaments white, hairy, bent in a semicircular manner to accommodate themselves to the cavity between the corolla and germ. Anthers oblong, orange coloured, ending in two double horns, bursting outwardly, for their whole length above the filaments, and not opening by pores as in Pyrola. Pollen white. Germ roundish, depressed, five angled, resting on a reddish, ten toothed, glandular ring. Style erect, straight. Stigma simple, moist. The fruit is a small, five celled capsule, invested with the calyx, which becomes large, round, and fleshy, having the appearance of a bright scarlet berry.

If the aroma or odour and also the taste of plants were susceptible of description in as definite language as their proportions and form, the sensible qualities of many vegetables might afford new grounds for generalizing and combining them together. The aromatic flavour of the Partridge berry, which cannot easily be mistaken by those who have once tasted it, may be recognised in a variety of other plants, whose botanical habits are very dissimilar. It exists very exactly in some of the other species of the same genus, particularly in *Gaultheria hispidula;* also in *Spiræa ulmaria* and the root of *Spiræa lobata.* It is particularly distinct in the bark of the Sweet birch, *Betula lenta,* one of our most useful and interesting trees.

This taste and odour reside in a volatile oil, which is easily separated by distillation. The essential oil of Gaultheria, which is often kept in our druggists' shops, is of a pale or greenish white colour and perfectly transparent. It is one of the heaviest of the volatile oils, and sinks rapidly in water if a sufficient quantity be added to overcome the repulsion of two heterogeneous fluids. Its taste is aromatic, sweet and highly pungent.

The oil appears to contain the chief medicinal virtue of the plant, since I know of no case in which the leaves, deprived of their aroma, have been employed for any purpose. They are nevertheless considerably astringent, and exhibit the usual evidences of this property when combined with preparations of iron.

The berries, or berry-like calyces, have a pulpy but rather dry consistence, and a strong flavour of the plant. They are esteemed by some persons, but are hardly palatable enough to be considered esculent. In the colder seasons they afford food to the partridges and some other wild animals.

The leaves, the essence and the oil of this plant are kept for use in the apothecaries' shops. An infusion of the leaves has been used to communicate an agreeable flavour to tea, also as a substitute for that article by people in the country. Some physicians have prescribed it medicinally as an emmenagogue, with success in cases attended with debility. The oil, though somewhat less pungent than those of peppermint and origanum, is employed for the same purposes. It shares with them the property of diminishing the sensibility of the nerve exposed by a carious tooth, when repeatedly applied. The essence, consisting of the volatile oil dissolved in alcohol or proof spirit, is antispasmodic and diaphoretic, and may be applied in all cases where warm or cordial stimulants are indicated. A tincture, formed by digesting the leaves in spirit, possesses the astringency as well as warmth of the plant, and has been usefully employed in diarrhæa.

A respectable physician of Boston informs me, that he has in various instances found the infusion of this plant very effectual in promoting the mammary secretion, when deficient; and even in restoring that important function after it had been for some time suspended. Whether the medicine has any specific influence of this sort, independent of the general state of the patient's health, I am not prepared to say.

BOTANICAL REFERENCES.

Gaultheria procumbens, LINN. *Sp. pl.*—MICHAUX, *Flor.* i. p. 249.—PURSH, i. 283.—NUTTALL, *Gen.* i. 263.—ANDREWS, *Bot. Repository*, t. 116.—WILLD. *Arb.* 123.—Vitis Idæa Canadensis Pyrolæ folio, TOURNEFORT, *Inst.* 608.—Anonyma pedunculis arcuatis, COLDEN, *Noveb.* 98.

MEDICAL REFERENCES.

KALM, *Amœnitates Academicæ*, iii. 14.—BART. *Coll.* i. 19.

PLATE XXII.

Fig. 1. *Gaultheria procumbens.*

Fig. 2. *The bractes or outer calyx.*

Fig. 3. *The true calyx.*

Fig. 4. *Stamen of the natural size.*

Fig. 5. *Anther magnified, the dark places shewing the mode of opening.*

Fig. 6. *Calyx and pistil.*

Fig. 7. *Fruit.*

Fig. 8. *Longitudinal section of the fruit.*

Fig. 9. *Transverse section of the capsule.*

PODOPHYLLUM PELTATUM.

May Apple.

PLATE XXIII.

THE *Podophyllum peltatum* or *May apple,* otherwise called *Mandrake* in this country, inhabits low shady situations from New England to Georgia. On the Atlantic coast I have never met with it farther north than Boston, yet in the interior of the country it has a more extensive range. From its large creeping roots, it has a great tendency to multiply, and is always found in beds of greater or less extent. Its flowering time is from March to May.

This plant is one of the *Ranunculaceæ* of Jussieu and *Rhœades* of Linnæus ; and is in the first order of the Class *Polyandria.*

Its generic character consists in a *calyx* of *three leaves ; from six to nine petals ; and a one-celled berry crowned with the stigma.* Only one spe-

Podophyllum peltatum

cies is at present known which strictly belongs to the genus.

The May apple has a jointed running root about half the size of the finger, by which it spreads extensively in rich grounds, where it gets introduced. The stem is about a foot in height, and invested at its base by the sheaths which covered it when in bud. It is smooth, round and erect, dividing at top into two round petioles from three to six inches long. Each petiole supports a large peltate, palmate leaf, smooth above, slightly pubescent beneath, deeply divided into about seven lobes, which are wedge shaped, two parted and toothed at the extremity. On the inside the leaf is cleft almost to the petiole. In barren stems which support but one leaf this does not take place, and the leaf is very perfectly peltate. In the fork of the stem is a solitary flower on a round nodding peduncle one or two inches long. Calyx of three oval, obtuse, concave leaves, cohering in the bud by their scarious margins, and breaking off at base when the flower expands. Petals from six to nine. Linnæus makes them nine in his generic character, but in this climate I have found them more frequently seven even in luxuriant specimens growing in very rich soil. They are obovate, obtuse, concave, smooth, white with slight transparent veins. Stamens shorter than the petals, curving upwards ; the anthers oblong, twice as long as their filaments. Germ oval, compressed, obscurely angular. Stigma nearly sessile, convex, its surface rendered irregular by numerous convolutions and folds. The flower is succeeded by a large ovate yellowish fruit, which is one celled, many seeded and crowned with the stigma. Its early period of ripening has given rise to the trivial name of *May apple.*

The dried root of the May apple is fragile and easily reduced to powder. It has a peculiar and rather unpleasant taste, but without much acrimony. When chewed for some time, it manifests a strong bitter taste. Both the tincture and decoction are intensely bitter. When water is added to the alcoholic solution the mixture becomes very gradually turbid, and at length opaque. On the other hand, alcohol disturbs both the infusion and decoction, especially the latter, in which it produces, after some time, a pearly whiteness. The trials I have made with it lead me to conclude that it contains a resin, a bitter extractive matter, fæcula and a slight proportion of a gummy substance.

The medicinal properties of the Podophyllum peltatum are those of a sure and active cathartic, in which character it deserves a high rank among our indigenous productions. We have hardly any native plant which answers better the common purposes of jalap, aloes and rhubarb, and which is more safe and mild in its operation. The root is the part to be employed, and should be given in substance in fine powder. I have commonly found twenty grains to operate with efficacy, and not to be attended with pain or inconvenience. In irritable stomachs it sometimes occasions nausea and vomiting, but this effect, as is well known, may ensue from any cathartic medicine. The late Professor Barton informs us, that although the root is an excellent cathartic, the leaves are poisonous, and the whole plant has something of a narcotic quality. Its botanical affinities would justify, *a priori*, a suspicion of this kind. In the various trials which I have made with it, I have not observed any such property in the root. The leaves I have never subjected to experiment for any purpose.

The fruit is acid and agreeable to the taste of many persons. It is sometimes called *wild lemons*, and is eaten with impunity.

The root is said by some physicians to be a medicine particularly suited to dropsy. It has also had the character in the Southern States of curing intermittent fever.

A physician in Albany informs me that the Shakers at Lebanon, N. Y. prepare an extract of the Podophyllum, which is much esteemed by medical practitioners as a mild cathartic. These people are well known to our druggists by the care and neatness with which they prepare a variety of medicines from native and naturalized pharmaceutical plants.

For medicinal use the root of the May apple is advised to be dug in the cold season, when vegetation is not active, viz. in the autumn and winter. At this part of the year the secretions of perennial plants are concentrated in their roots, and the same weight of their substance is less diluted with the watery or ascending sap, than it is at any other period. This constitutes a reason why the roots of all perennial plants should, as far as practicable, be taken up during the cold season. But from what I have been able to observe, the difference of their virtue in different months is much less than is commonly supposed. I never knew a medicinal plant whose efficacy was destroyed in consequence of being taken up even at midsummer, although it may be in some degree lessened. It is probable that those roots which constitute staple articles of commerce, as ipecac, gentian, rhubarb, &c. are gathered indiscriminately for exportation at all seasons when they are to be found. Being collected by savages or by ignorant persons, who seek for them in their native wilds, and who are not much interested in their future efficacy; it is probable they would be gathered in greatest quantities when their vegetation was most luxuriant, because at this time their shoots and tops would be most conspicuous. We know this to be the case with our Ginseng, Spigelia, Snake root, &c. which form considerable articles of exportation, and which it would be difficult to find at any other than the vegetating season.*

* *Annual plants* should be gathered at the time when their vegetation is most vigorous, which is generally from the time they begin to flower, until the leaves begin to change. The leaves contain the greatest activity in most annual plants employed for medicine, while the root is a comparatively insignificant part, being small, woody and fibrous. Thus the leaves of Stramonium and Tobacco are much more active than the root.

Biennial plants should, in most instances, be gathered in the second season of their growth, and about the time of flowering The leaves of these plants also contain their medicinal activity, as in Hemlock and Henbane. The roots are medicinal, but usually in a less degree. In some aromatic biennials, the seeds are the most important part of the plant.

BOTANICAL REFERENCES.

Podophyllum peltatum, Linn. *Sp. pl.*—Michaux, *Flora*, i. 309.—Pursh, ii. 366.—Lamarck, *Illust. gen.*—Trew, *Ehret. t.* 29.—Anapodophyllum Canadense, Catesby, *Car.* i. 24.—Aconitifolius humilis &c. Mentz. *pug. t.* 11.

MEDICAL REFERENCES.

Schœpf, 86.—B. S. Barton, *edit. of Cullen*, 375.—Thacher, *Disp.* 307.—Chapman, *Mat. Med.* 209.

PLATE XXIII.

Fig. 1. *Podophyllum peltatum.*

Fig. 2. *Calyx.*

Fig. 3. *Stamens.*

Fig. 4. *Germ and stigma.*

Fig. 5. *Fruit.*

Pl. XXIV.

ICTODES FŒTIDUS.

Skunk Cabbage.

PLATE XXIV.

THIS is one of our most noticeable plants, both from the frequency of its occurrence and the peculiarity of its sensible properties. Scarcely a swamp or meadow is found in the middle and northern parts of the United States in which this vegetable may not be discovered at a distance, especially in the spring season, by its large tufts of rank, crowded leaves. Its singular flowers are among the first which break from the ground, after the rigours of winter, appearing in different latitudes, from February to April. The vegetation is rapid, so that in most instances the fruit is ripe and the leaves wholly decayed before the end of August. From this precocity of the plant together with the depth to which the roots penetrate the earth, it seems calculated to bear the cold of high latitudes. I have found the flowers a second time formed, and shooting from the ground in November. The strong and unpleasant odour which every part of the plant emits on being broken, and which is precisely similar to that of the *Viverra mephitis*; has given it by an almost common consent, in every part of the country, the appellations of *Skunk weed* and *Skunk cabbage*.

The structure of this singular vegetable has caused it successively to be assigned to the genera *Arum*, *Dracontium* and *Pothos*, with none of which it fully agrees. Of the Aroideæ, to which it is related, it approaches most nearly in its flower to *Pothos*; while its fruit has more affinity to *Orontium*. The Rev. Dr. Cutler many years ago, in the Transactions of the American Academy, pointed out the distinctive characters of this plant, and pronounced it a new genus. No name, however, substantiated by a character, has to my knowledge been given it, in any botanical work, except the name of *Symplocarpus*, a term lately adopted by some American botanists on the alleged authority of Mr. Salisbury. As this name by its etymology implies a resemblance of the fruit to *Symplocos*, a genus with which the plant has not the least affinity; it appears to me inadmissible. Although I am averse to multiply the confusion of synonyms, with which our science is already too much burdened, yet in the present instance an appropriate name, which should not be at variance with the character of the plant, appeared to be required. With the advice of the venerable Dr. Cutler, I have translated, as nearly as possible, the common English appellation for the plant. The name *Ictodes* from ἴκτις, *viverra*, and οζω, *oleo;* is sufficiently expressive of the property from which its common name is derived.

The genus Ictodes has for its character a *hooded spathe, spadix covered with perfect flowers, calyx with four segments, petals none, style pyramidal, seeds immersed in the spadix.* Only the present plant can be assigned to this genus. It belongs to *Tetrandria, monogynia;* and is found among the *Piperitæ* of Linnæus and *Aroideæ* of Jussieu. The root is large and abrupt, with numerous, crowded, fleshy fibres. The spathe which emerges from the ground some time before the leaves, is ovate, swelling, various in width, cucullate, spotted and sometimes nearly covered with dull brownish purple, the top acuminate and incurved, the edges infolded, auriculate at base, and at length coalescing. Within this is the oval spadix, on a short peduncle, covered with perfect tetrandrous flowers, and of the same colour with the spathe. Calyx leaves four, fleshy, wedge shaped, truncate, the top and edges inflected, the whole crowded together so as to form a compact covering for the spadix. Stamens four, opposite the calyx leaves, with subulate filaments equal in length to the calyx, and oblong four celled anthers. Style four sided, tapering; stigma minute, pubescent; germ roundish, concealed within the spadix. After the spathe decays, the spadix continues to grow, and with it every part of the flowers except the anthers. When the fruit is ripe, the spadix has attained many times its original dimensions, while the calyx, filaments and style are larger, very prominent and separated from each other. Within the spadix at the base of each style is a round, fleshy seed, as large as a pea, white, tinged with green and purple, invested with a separate membranous coat, and with a prominent corculum situated in a depression at top.

The leaves which spring up some time after the flowers are numerous, large and crowded, oblong heart shaped, acute, smooth, with numerous fleshy veins of a paler colour. They spring from the root on long petioles, hollowed in front, and furnished with large oblong sheaths. They continue to increase in size for a month or two after the flowering period is past.

Mr. Nuttall, who has observed the germination of this plant, informs us that the seed does not appear to possesss any other cotyledon, than a sheathing stipule, similar to that which is afterwards produced in the plant. The principle bulk of the seed is formed by what he considers a vitellus, having the embryo exactly resembling the future plant, situated in an umbilical depression at its top. The attachment of this body to the embryo is at first by a minute *funiculus*, which enlarges and becomes more distinct during the progress of germination; but the most singular circumstance respecting it is the length of time for which it continues attached to the growing plant, apparently inert at the base of the candex for twelve or even eighteen months.

The offensive and powerful odour which characterizes this plant is not peculiar to it. The fruit of some of the North American currants, and particularly *Ribes rigens* of Michaux, a species often met with on the high mountains of the Eastern States; emits when bruised a scent exactly similar to this vegetable.

The odour of the Ictodes resides in a principle which is extremely volatile. I have not been able to separate it by distillation from any part of the plant, the decoction and the distilled water being in my experiments but slightly impregnated with its sensible character. Alcohol, digested on the plant, retains its odour for a time, but this is soon dissipated by exposure to the air.

An acrid principle exists in the root even when perfectly dry, producing an effect like that of the Arum and Ranunculi. When chewed in the mouth, the root is slow in manifesting its peculiar taste; but after some moments, a pricking sensation is felt, which soon amounts to a disagreeable smarting, and continues for some time. This acrimony is readily dissipated by heat. The decoction retains none of it. The distilled water is impregnated with it, if the process be carefully conducted, but loses it on standing a short time.

A resinous substance is dislodged from the alcoholic solution of the root by the addition of water, the solution becoming moderately turbid. A gummy or mucous principle is also present, and fills the mouth with mucilage when the root is chewed. It is separated from the decoction in small *flocculi* when alcohol is added.

The spadix consists of a fleshy cellular substance, which shrinks very much in drying. The seeds when dry are reduced to half their former size, and in this state they have a tough waxy consistence and an animal odour. They contain fixed oil in abundance, which is easily forced out from them by expression. Their principal bulk appears to be albumen, and when reduced to powder they are less easily soluble in boiling water, than grains which are less oleaginous. They burn with an oily smoke, leaving behind a large coal.

The sensible properties of the Ictodes having a strong affinity with those of assafœtida and the other fœtid gums, practitioners have been led to expect from it a similar antispasmodic power. Experience has justified these expectations in a variety of disorders of the spasmodic and nervous kind. The Rev. Dr. Cutler of Massachusetts was the first who recommended its use in asthmatic cases. In his account of indigenous American vegetables, he tells us that the roots dried and powdered form an excellent remedy in asthma, and often give relief when other means prove ineffectual. It may be given, he says, with safety to children as well as adults; to the former in doses of four, five or six grains, and to the latter in doses of twenty grains and upward. In a private letter he states, that he made use of it in his own case of asthma for several years, and generally found relief. In the winter he used the dried root in powder, and in summer, the fresh grated root. It continued to afford more relief than any other remedy, so long as the paroxysms remained under the influence of any medicine. Since the recommendation of Dr. Cutler, many country physicians have employed the root in asthma, catarrh and chronic coughs, with evident benefit. A number of cases have fallen under my own observation of the catarrh affections of old people, in which a syrup prepared from the root in substance has alleviated and removed the complaint. Dr. Thacher informs us on various authorities, that the powdered root has given immediate relief in hysteric paroxysm, that it has affected the cure of dropsy, and that rheumatic patients have found great benefit from its use. Its strong and penetrating acrimony would lead us, *a priori*, to expect advantage from it in these complaints. Even in the more formidable disease of epilepsy, it has appeared to do good.

Some caution, however, is requisite in its management, as serious inconvenience may ensue from an over dose. In delicate stomachs I have found it frequently to occasion vomiting even in a small quantity. In several cases of gastrodynia where it was given with a view to its antispasmodic effect, it was ejected from the stomach more speedily than common cathartic medicines. I have known it in a dose of thirty grains to bring on not only vomiting, but headach, vertigo and temporary blindness. Other practitioners have given it in larger quantities without any evil of this kind, but I think such an exemption must be attributed to the age and deteriorated quality of the root. Its active ingredients being more or less volatile, it must necessarily be impaired in strength by long keeping, especially in a pulverized state.

To insure a tolerably uniform activity of this medicine, the root should be kept in dried slices and not reduced to powder until it is wanted for use. It may then be taken in pills or mixed with syrup in doses of from ten to twenty grains. These may in most instances be repeated three times a day.

BOTANICAL REFERENCES.

Arum Americanum, CATESBY, *Car.* ii. *t.* 71.—Dracontium fœtidum, LIN. *Syst. pl.*—WILLD. ii. 288.—Pothos fœtida, MICHAUX, *Amer.* ii. 186.—PURSH, ii. 398.—*Bot. Mag.* 836.—Symplocarpus fœtida, NUTTALL, *genera,* i. 105.

MEDICAL REFERENCES.

CUTLER, *Trans. Amer. Acad.* i. 407. THACHER, *Dispensatory,* 150.

PLATE XXIV.

Fig. 1. *Ictodes fœtidus in flower, the spathe inclosing the spadix.*

Fig. 2. *The spadix taken out of the spathe.*

Fig. 3. *The leaves, stalks, &c.*

Fig. 4. *The spadix in fruit, one quarter being cut away to show the seeds.*

Fig. 5. *A flower magnified.*

Fig. 6. *The same opened.*

Fig. 7. *Petal and Stamen.*

Fig. 8. *Style.*

Fig. 9, 10. *Seeds.*

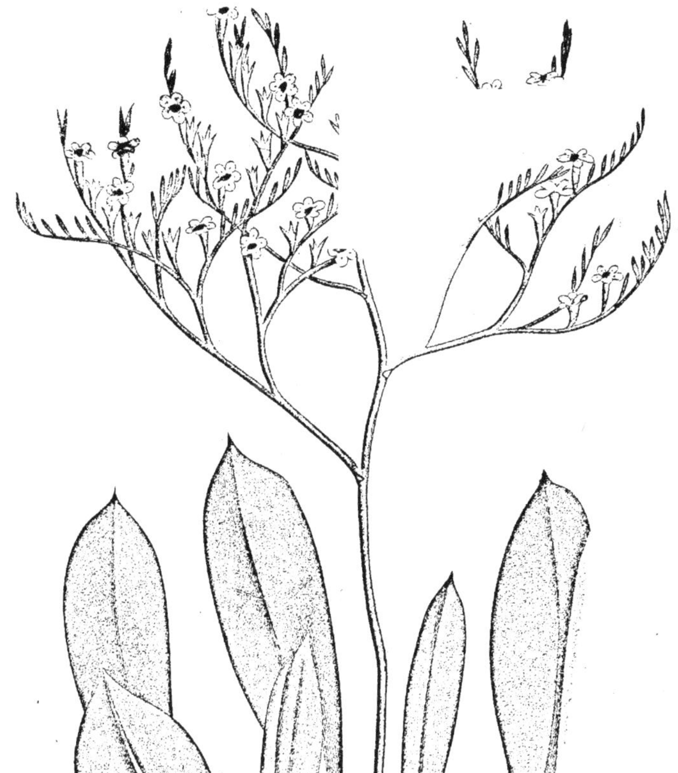

Statice Caroliniana

STATICE CAROLINIANA.

Marsh Rosemary.

PLATE XXV.

The class of vegetables, denominated maritime, or sea shore plants, are constituted to occupy extensive tracts of ground, which, from their impregnation with sea salt, are incapable of sustaining the life and growth of other species. The muriate of soda, if poured at the roots of the most vigorous plants belonging to a fresh soil, will often destroy them in a short time. Few forest trees of the temperate zones can grow in marshes where their roots are wholly exposed to the access of salt water. Yet such is the wise arrangement of nature, that this substance, which proves a poison to most vegetables, is converted into the food and necessary stimulus of the rest. Maritime plants flourish alike in places visited by the tide, and those impregnated by the salt springs of the interior. The degree in which they require the presence of the mineral is various, some growing upon the beach, where the earth is saturated with salt, and others at the extreme edge of marshes, where the impregnation is much less powerful. With a few exceptions, they cannot long be cultivated in fresh earth, but soon decay when removed from their native marshes.

Maritime plants derive a peculiar character from their place of growth, which distinguishes them even when dry from other vegetables. The salt with which they are impregnated crystallizes on their surface in dry weather; and deliquesces so as to render them damp and supple, when the atmosphere is moist. These plants are troublesome in an herbarium from the facility with which they contract moisture from the atmosphere, and communicate it to the adjacent papers. The hay cut upon salt marshes often becomes extremely damp, and would be entirely spoiled, were it not for the antiseptic and preservative quality of the salt. The barilla of commerce is obtained by the combustion of maritime vegetables.

Many of these plants are thick and fleshy in their mode of growth, and differ remarkably in this respect from their co-species on dry ground. This is particularly seen in Arenaria, Gerardia, Chenopodium, &c.

The vegetable which is the subject of this article is exclusively a maritime plant. It is one of the few ornamental species in our salt marshes, and is very conspicuous for its purple tops appearing among the grass in all the summer months. It varies from a few inches, to a foot and more in height.

This species has generally been considered a variety of the *Statice limonium*, which is a common plant in the salt marshes of Europe. Indeed, several of the maritime species of this genus approach each other so closely in their characters, that they have been considered the same by able botanists. The American plant, to which the name of *Caroliniana* was given by Walter in his Flora of Carolina, is distinguished from the European principally by its smaller flowers and plain or flat leaves. From the *Statice Gmelini*, an Asiatic species, it differs apparently still less in its general form.

The genus *Statice* belongs to the class *Pentandria* and order *Pentagynia*. Its natural orders are *Aggregatæ* of Linnæus and *Plumbagines* of Jussieu. It is characterized by *a calyx monophyllous, pluited* and *scarious. Petals five with the stamens inserted in their claws. Seed one, invested with the calyx.* The species *Caroliniana* has its *scape round* and *panicled; its leaves obovate-lanceolate, smooth, obtuse, mucronated, and flat on the margin.*

The root of this plant is perennial, large, fleshy, fusiform or branched. Several tufts of leaves and scapes are often produced from the same root. The leaves are narrow-obovate, supported by long petioles, smooth, veinless, obtuse, mucronated by the prolongation of the middle rib, level and flat on the margin, in which respect they differ from S. limonium, which is undulated. Scape round, smooth, furnished with a few scales, flexuous at top, giving off numerous branches, which end in spikes of flowers ; the whole forming a large panicle. The base of each branch and flower is supported by an ovate, mucronated scale. The flowers are alternate, erect, consequently one sided in the horizontal branches ; mostly in pairs, but appearing single from one expanding before the other. They grow on a short, forked peduncle, which is concealed by several sheathing scales, part of which are common to the two, and part peculiar to the upper one. The calyx is funnel shaped, five angled, the angles ciliate and ending in long acute teeth with sometimes, not always, minute intermediate teeth. The upper part of the calyx is scarious and of a pink colour. Petals spatulate, obtuse, longer than the calyx, pale bluish purple. Stamens inserted in the claws of the petals, anthers heart shaped. Germ small, obovate, with five ascending styles shorter than the stamens. Seed oblong, invested with the persistent calyx.

The root, which is the officinal part of the Marsh Rosemary, is one of the most intense and powerful astringents in the vegetable materia medica. It communicates to the mouth an highly austere and astringent taste, combined with a good deal of bitterness. Few vegetable substances, when chemically treated, give more distinct and copious evidence of the presence of both tannin and gallic acid. The sulphate of iron strikes a fine purple colour with the solution, and soon deposits a precipitate, which, on exposure to the air, becomes of an inky blackness. Gelatin also throws down a copious, whitish, insoluble precipitate. Resin hardly exists in this root, nor any thing else exclusively soluble in alcohol. The impregnation with sea salt is readily made obvious.

Dr. Mott, Professor of Surgery in the University of New York, has published an interesting and valuable investigation of the properties of this plant in 1806. He informs us that the astringency, indicated by the sulphate of iron, was greater in the tincture than in the infusion under experiments precisely similar; from which it may be inferred, that alcohol is a better solvent for this root than water. He also found the cold infusion more powerful than the hot, a circumstance probably to be accounted for by the escape of a part of the gallic acid by evaporation. The astringency was found fully equal to that of galls, and ink made from equal quantities of the two, similarly treated, was equal in blackness.

The Statice Caroliniana possesses much medicinal reputation as an astringent, and large quantities of it are annually consumed in different parts of the United States. In Boston it is regularly kept by the druggists, and larger quantities are sold, than of almost any indigenous article. It is principally sought for as a topical remedy in aphthæ and other ulcerative affections of the mouth and fauces. From its astringent and antiseptic quality, it is peculiarly suited to correct the state of these local maladies, and its application is commonly followed with success. It is much better suited to such complaints than the Coptis trifolia or gold thread, with which it is frequently combined, and which is only a tonic bitter without astringency.

Dr. Baylies of Dighton, Mass. employed a decoction of the root, both internally and externally, in the Cynanche maligna, a disease which has at times been epidemic and very destructive in different parts of our country. It proved very successful not only under his own observation, but under that of other physicians in this dangerous complaint.

Dr. Mott informs us, that in the chronic stages of dysentery, after the inflammatory diathesis, great tenesmus, &c. are removed; a strong decoction of the root has restored patients to health, after various tonics and astringents had been used to no effect.

BOTANICAL REFERENCES.

Statice Caroliniana, WALTER, *Flora Car.* 118.—PURSH, i. 212.—NUTTALL, i. 206.—Statice limonium, MUHLENBERG, *Catalogue*, 33.—ELLIOTT, *Carolina*, i. 374.

MEDICAL REFERENCES.

MOTT, *Inaugural Dissertation.*—THACHER, *Disp.* 345.—BAYLIES, *Papers of the Mass. Med. Society*, vol. i.

PLATE XXV.

Fig. 1. *Statice Caroliniana.*

Fig. 2. *A flower magnified.*

Fig. 3. *Calyx ditto.*

Fig. 4. *A petal and stamen ditto.*

Fig. 5. *Germ and styles ditto.*

XXVI.

Asclepias tuberosa

ASCLEPIAS TUBEROSA.

Butterfly Weed.

PLATE XXVI.

FEW genera are more curious and intricate in their structure, than that to which our present article belongs. The plants which constitute the family of Asclepias are so peculiar in their habit, that they are easily recognized even by the inexperienced botanist, while their minute structure is so complicated, as to require not a little attention for its perfect development. This fine race of plants are so abundant in the United States, that every month of the summer season presents us a number of beautiful species. By far the most rich and gaudy of these in appearance is the Asclepias tuberosa, known by the vulgar names of *Butterfly weed* and *Pleurisy root*, and found in dry, sandy soils, pine woods, &c. from Massachusetts to Georgia. It is the Asclepias decumbens of Walter.

This genus has a *five parted calyx; a five parted reflexed corolla; a nectary of five erect, cucullate leaves, each producing an inflected horn from its cavity; stamens united, with ten pollen masses hanging by pairs in their cavities.* The species *tuberosa* is *hairy, its leaves alternate, oblong-lanceolate; its branches cymose.*

Class *Pentandria*, order *Digynia.* Natural orders *Contortæ*, L. *Apocineæ*, Juss.

The root of this plant is large, fleshy, branching, and often somewhat fusiform. It is only by comparison with the other species that it can be called tuberous. The stems are numerous, growing in bunches from the root. They are erect, ascending or procumbent, round, hairy, green or red. Leaves scattered, the lower ones pedunculated, the upper ones sessile. They are narrow, oblong, hairy, obtuse at base, waved on the edge, and in the old plants sometimes revolute. The stem usually divides at top into from two to four branches, which give off crowded umbels from their upper side. The involucrum consists of numerous, short, subulate leafets. Flowers numerous, erect, of a beautifully bright orange colour. Calyx much smaller than the corolla, five parted, the segments subulate, reflexed and concealed by the corolla. Corolla five parted, reflexed, the segments oblong. The nectary or stamineal crown is formed of five erect, cucullate leaves or cups, with an oblique mouth, having a small, incurved, acute horn proceeding from the base of the cavity of each and meeting at the centre of the flower. The mass of stamens is a tough, horny, somewhat pyramidal substance, separable into five anthers. Each of these is bordered by membranous, reflected edges contiguous to those of the next, and terminated by a membranous, reflected summit. Internally they have two cells. The pollen forms ten distinct, yellowish, transparent bodies, of a flat and spatulate form, ending in curved filaments, which unite them by pairs to a minute dark tubercle at top. Each pair is suspended in the cells of two adjoining anthers, so that if a needle be inserted between the membranous edges of two anthers and forced out at top, it carries with it a pair of the pollen masses. Pistils two, completely concealed within the mass of anthers. Germs ovate, with erect styles. The fruit, as in other species, is an erect lanceolate follicle on a sigmoid peduncle. In this it is green, with a reddish tinge and downy. Seeds ovate, flat, margined, connected to the receptacle by long silken hairs. Receptacle longitudinal, loose, chaffy.

The down or silk of the seeds, in this and other species, furnishes an admirable mechanism for their dissemination. When the seeds are liberated by the bursting of the follicle which contains them, the silken fibres immediately expand so as to form a sort of globe of branching and highly attenuated rays, with the seed suspended at its centre. In this state they are elevated by the wind to an indefinite height, and carried forward with a voyage like that of a balloon, until some obstacle intercepts their flight, or rain precipitates them to the ground.

The down of different species of Asclepias is susceptible of application to various useful and ornamental purposes. If the fibre were sufficiently long to admit of its being woven or spun, it would approach more closely to silk in its gloss and texture, than any vegetable product we possess. As it is, it has been substituted for fur, in the manufacture of hats, and for feathers in beds and cushions. When attached by its ends to any woven fabric, this down forms a beautiful imitation of the finest and softest fur skins, and is applicable to various purposes of dress. The Asclepias Syriaca, from its frequency and the large size of its pods, has been most frequently employed for the foregoing purposes. [*Note* A.]

The root of the Butterfly weed when dry is brittle and easily reduced to powder. Its taste is moderately bitter, but not otherwise unpleasant. Its most abundant soluble portions are a bitter extractive matter and fæcula. No evidence of astringency is afforded on adding solutions of isinglass or copperas, and hardly any traces of resin on adding water to alcohol digested on the root. The decoction afforded a flaky precipitate to alcohol, when the infusion did not. Boiling water may be considered the proper menstruum for this plant.

This fine vegetable is eminently intitled to the attention of physicians as an expectorant and diaphoretic. It produces effects of this kind with great gentleness, and without the heating tendency which accompanies many vegetable sudorifics. It has been long employed by practitioners in the Southern States in pulmonary complaints, particularly in catarrh, pneumonia and pleurisy, and has acquired much confidence for the relief of these maladies. It appears to be an expectorant peculiarly suited to the advanced stages of pulmonary inflammation, after depletion has been carried to the requisite extent. Dr. Parker of Virginia, as cited by Dr. Thacher, having been in the habit of employing this root for twenty five years, considers it as possessing a peculiar and almost specific quality of acting upon the organs of respiration, promoting suppressed expectoration, and relieving the breathing of pleuritic patients in the most advanced stage of the disease.

Dr. Chapman, Professor of medicine in Philadelphia, informs us that his experience with this medicine is sufficient to enable him to speak with confidence of its powers. As a diaphoretic he thinks it is distinguished by great certainty and permanency of operation, and has this estimable property, that it produces its effects without increasing much the force of the circulation, raising the temperature of the surface, or creating inquietude and restlessness. On these accounts it is well suited to excite perspiration in the forming states of most of the inflammatory diseases of winter, and is not less useful in the same cases at a more advanced period, after the reduction of action by bleeding, &c. The common notion of its having a peculiar efficacy in pleurisy, he is inclined to think is not without foundation. Certain it is, says he, that it very much relieves the oppression of the chest in recent catarrh, and is unquestionably an expectorant in the protracted pneumonies.

As far as my own observation with this plant extends, I am persuaded of its usefulness in various complaints. It appears to exert a mild tonic effect, as well as a stimulant power on the excretories. Like other vegetable bitters, if given in large quantities, especially in infusion and decoction, it operates on the alimentary canal, though its efficacy in this respect is not sufficient to entitle it to rank among active cathartics. I am satisfied of its utility as an expectorant medicine, and have seen no inconsiderable benefit arise from its use as a palliative in phthisis pulmonalis. Among other instances may be cited that of a young physician in this town, who died two years since of pulmonary consumption. He made great use of the decoction of this root, and persevered in it a long time from choice, finding that it facilitated expectoration and relieved the dyspnœa and pain in the chest, more than any other medicine.

The best mode of administering the Asclepias is in decoction or in substance. A teacup full of the strong decoction, or from twenty to thirty grains of the powder, may be given in pulmonary complaints several times in a day. In most cases after the inflammatory diathesis is in some degree subsided, it may be freely repeated as long as it agrees with the stomach and bowels.

BOTANICAL BEFERENCES.

Asclepias tuberosa, LIN. *Sp. pl.*—PURSH, i. 183.—MICHAUX, i. 117.—ELLIOTT, *Car.* i. 325.—Asclepias decumbens, *a variety*, LIN. PURSH, &c.—Apocynum Novæ Angliæ hirsutum radice tuberosa, floribus aurantiacis, HERMAN, *Hort.* 646. *t.* 647.—DILLENIUS, *Elth.* 35, *t.* 30, *f.* 34.

MEDICAL REFERENCES.

B. S. BARTON, *Collections*, 48.—THACHER, *Disp.* 154.—CHAPMAN, *Therapeutics and Mat. Med.* i. 346.

PLATE XXVI.

Fig. 1. *Asclepias tuberosa.*

Fig. 2. *A flower.*

Fig. 3. *A flower dissected, showing the mass of anthers, and one nectary with its horn.*

Fig. 4. *Magnified section of the mass of anthers, showing the situation of the pistils inside, &c. A pair of pollen masses is drawn out at the top.*

Fig. 5. *Pistils magnified, and calyx.*

3 

Magnolia glauca

4 

MAGNOLIA GLAUCA.

Small Magnolia.

PLATE XXVII.

Of the splendid family of trees known by the name of Magnolia, the American continent has many species. Taken collectively they furnish perhaps the most elegant assemblage produced in the forests of the temperate zone. They are distinguished by their rich, smooth foliage, large fragrant flowers, and aromatic bark. Some of them are trees of very exalted stature, taking rank with the highest tenants of the woods. The present species is more humble than the rest in its growth, yet more interesting in some of its other properties.

The Magnolia glauca has the most extensive range, especially near the sea board, of any of the species of its family. Its most northern boundary appears to be in a sheltered swamp in Manchester, Cape Ann, about thirty miles north of Boston. It here attains to but small size, and is frequently killed to the ground by severe winters. It is common in the Middle and Southern States, and Michaux informs us, that it is one of the most abundant trees in the morasses of Florida and Lower Louisiana. According to this author however, it is not usually met with far in the interior, or to the west of the mountains. Its common names are various, and change with almost every district. In Massachusetts it has no other name than *Magnolia*; in the Middle States it is called *Swamp sassafras* and *Beaver tree*; while in the Southern States it is denominated *Sweet bay* and *White bay*. It is naturally a tenant of deep boggy swamps, and is somewhat irregular in its growth. It acquires more symmetry of form when cultivated in an upland soil, although its transplantation is difficult. To insure it sussessful cultivation in a dry soil, the tree should be raised from the seed.

This tree begins to flower in different parts of the United States in May, June and July. The flowers are highly fragrant, and may be perceived by their perfume at a considerable distance. A few of them shut up in a room over night communicate to the air a heavy and almost insupportable odour.

The Magnolias are found in the class *Polyandria* and order *Polygynia*; the *Coadunatæ* of Linnæus and *Magnoliæ* of Jussieu.

This genus has a *calyx of three leaves, a corolla of six petals or more; capsules two-valved, imbricated, forming a cone; seeds berried, pendulous.*

The present species has *oval leaves, glaucous underneath*; and *obovate petals, narrowed at base.*

The bark of the young twigs is of a bright, smooth green, with rings at the insertion and scars of the leaves. The leaves are scattered, petioled, regularly elliptical, entire, and glabrous. Their under side, except the midrib, is of a beautifully pale, glaucous colour, by which the tree may be distinguished at a distance. When young, this surface is covered with a silken pubescence. Flowers solitary, terminal, on a short, incrassated peduncle. Calyx of three spatulate, obtuse, concave segments. Corolla of from eight to fourteen obovate, obtuse, concave petals, contracted at their base. The stamens are very numerous and inserted in common with the petals on the sides of a conical receptacle. Filaments very short; anthers linear, mucronated, two-celled, opening inwardly. Germs oval, collected into a cone, each one divided by a furrow and tipt with a brownish, linear, recurved style. The fruit is a cone, consisting of imbricated cells, which open longitudinally for the escape of the seed. The seeds are obovate, scarlet, connected to the cone by a thread, which suspends them some time after they have fallen out.

The bark of the Magnolia glauca has a bitter taste, combined with a strong aromatic pungency, which approaches that of *Sassafras* and of the *Acorus calamus*. The aroma resides in a volatile portion, which is probably an essential oil, or a variety of camphor. It is lost from the bark in the dry state, after it has been kept some time. Water distilled from the green bark has its peculiar flavour with an empyreumatic smell. No oil appears on the surface, when the experiment is conducted in the small way. The dried bark affords a little resin, and more of a bitter extractive substance. Chalybeate tests produce a very slight darkening of the green colour of the decoction, but gelatin occasions no change. This might be anticipated from the little taste of astringency in the bark.

As a medicinal article, the Magnolia is to be considered an aromatic tonic, approaching in its character to Cascarilla, Canella, and articles of their class. Considered simply in regard to its tonic powers, it is probably of a secondary order, though from the additional properties which it possesses of a warm stimulant and diaphoretic is found useful in certain disorders. Chronic rheumatism is one of the diseases in which it exhibits most efficacy. Not only the bark, but the seeds and cones which are strongly imbued with the sensible qualities of the tree, are employed in tincture with very good success in this disease.

In intermittent and remittent fevers the Magnolia is one of the many tonics which have been resorted to for cure by the inhabitants of the marshy countries where they prevail. Sufficient testimony has been given in favour of the bark of this tree, to warrant a belief that it is fully adequate to the removal of fever and ague, when administered like the Cinchona, in liberal quantities between the paroxysms. In the more continuous forms of fever of the typhoid type, it has also received the commendations of physicians.

Several other species of Magnolia resemble the present very closely in their sensible properties, and as far as experiments have been tried, they are similar in their medicinal effects. In order to secure the whole efficacy residing in these trees, a tincture should be made from the bark or cones while green or very recently dried, before their more volatile parts have escaped.

linear, recurved style. The fruit is a cone, consisting of imbricated cells, which open longitudinally for the escape of the seed. The seeds are obovate, scarlet, connected to the cone by a thread, which suspends them some time after they have fallen out.

The bark of the Magnolia glauca has a bitter taste, combined with a strong aromatic pungency, which approaches that of *Sassafras* and of the *Acorus calamus*. The aroma resides in a volatile portion, which is probably an essential oil, or a variety of camphor. It is lost from the bark in the dry state, after it has been kept some time. Water distilled from the green bark has its peculiar flavour with an empyreumatic smell. No oil appears on the surface, when the experiment is conducted in the small way. The dried bark affords a little resin, and more of a bitter extractive substance. Chalybeate tests produce a very slight darkening of the green colour of the decoction, but gelatin occasions no change. This might be anticipated from the little taste of astringency in the bark.

As a medicinal article, the Magnolia is to be considered an aromatic tonic, approaching in its character to Cascarilla, Canella, and articles of their class. Considered simply in regard to its tonic powers, it is probably of a secondary order, though from the additional properties which it possesses of a warm stimulant and diaphoretic is found useful in certain disorders. Chronic rheumatism is one of the diseases in which it exhibits most efficacy. Not only the bark, but the seeds and cones which are strongly imbued with the sensible qualities of the tree, are employed in tincture with very good success in this disease.

In intermittent and remittent fevers the Magnolia is one of the many tonics which have been resorted to for cure by the inhabitants of the marshy countries where they prevail. Sufficient testimony has been given in favour of the bark of this tree, to warrant a belief that it is fully adequate to the removal of fever and ague, when administered like the Cinchona, in liberal quantities between the paroxysms. In the more continuous forms of fever of the typhoid type, it has also received the commendations of physicians.

Several other species of Magnolia resemble the present very closely in their sensible properties, and as far as experiments have been tried, they are similar in their medicinal effects. In order to secure the whole efficacy residing in these trees, a tincture should be made from the bark or cones while green or very recently dried, before their more volatile parts have escaped.

BOTANICAL REFERENCES.

Magnolia glauca, Lin. *Sp. pl.*—Michaux, i. 323.—Pursh, ii. 381.—Michaux, fil. *Arb. forest*, iii. 77.—Magnolia lauri folio subtus albicante. Catesby, *Car.* i. *t.* 39.—Trew, *sel. t.* 9.—Dillenius, *Hort.* 207. *t.* 168, *f.* 205.—Laurus tulipifera &c.—Raius, *hist.* 1690.

MEDICAL REFERENCES.

Kalm, *Travels*, i. 205.—Marshall, *Arbustum*, 83.—Humphries, *Med. Commentaries*, vol. xviii.—Bart. *Coll.* 46.—Price, *Inaugural Diss. Philad.* 1812.

PLATE XXVII.

Fig. 1. *A flowering branch of Magnolia glauca.*

Fig. 2. *The fruit and seeds.*

Fig. 3. *Stamen magnified.*

Fig. 4. *A germ and style ditto.*

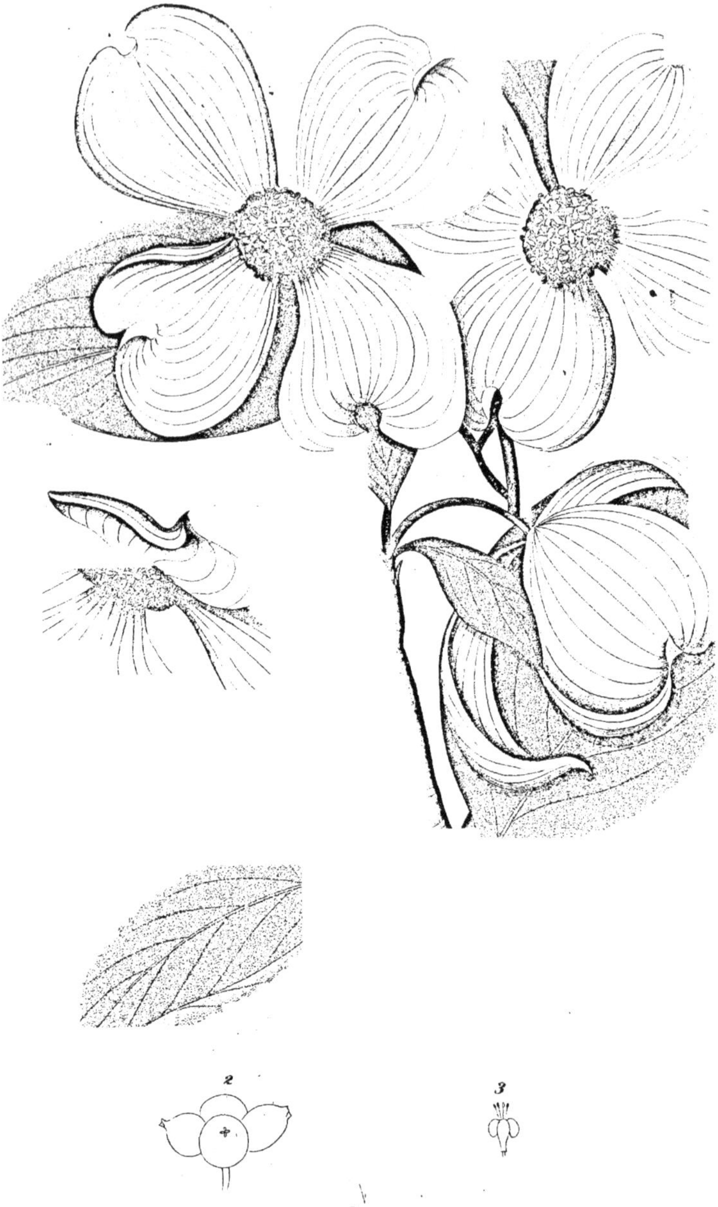

1 *Cornus florida*

CORNUS FLORIDA.

Dogwood.

PLATE XXVIII.

THE family of Cornels, if surveyed by other eyes than those of botanists, is remarkable for the difference of growth and appearance of its various species. Many of them are shrubs; a few attain to the stature of trees, while some are so humble in their growth as to be deemed hardly more than herbaceous. A part have their flowers surrounded with a fine white involucrum, many times exceeding the whole bunch in magnitude; while others present their naked cymes unadorned by any investment. To the botanical observer they all exhibit a close affinity and resemblance to each other; which is seen in the form and anatomical texture of their leaves, the structure of their flowers and the appearance of their fruit.

10

The *Cornus florida*, or flowering Dogwood, is the largest and most splendid of its genus, and is one of the chief ornaments of our forests. As a tree it is rather below the middle stature, not usually reaching the height of more than twenty or thirty feet. It is however among the most conspicuous objects in the forests, in the months of April, May and June, according to its latitude, being then covered with a profusion of its large and elegant flowers. In Massachusetts, especially about Boston, it is not a common tree, only scattered individuals appearing here and there in the woods. In the Middle States it is extremely common, especially in moist woods. Michaux informs us, that in the Carolinas, Georgia and the Floridas it is found only on the borders of swamps, and never in the pine barrens, where the soil is too dry and sandy to sustain its vegetation. It is also not very common in the most fertile parts of the Western States, being chiefly found where the soil is of secondary quality.*

* Mr. William Bartram, in his travels in Georgia and Florida, gives the following account of the appearance of this tree near the banks of the Alabama river. "We now entered a very remarkable grove of Dogwood trees, (*Cornus florida*,) which continued nine or ten miles unalterable, except here and there a towering *Magnolia grandiflora*. The land on which they stand is an exact level; the surface a shallow, loose, black mould, on a stratum of stiff, yellowish clay. These trees

The genus *Cornus* is characterized by the following marks. *Petals four, superior; involucrum of four leaves, or wanting; drupe with a two-celled nut.* The species *florida* is *arboreous, with its flowers in heads surrounded by an involucrum of obovate leaves with recurved points.*

Class *Tetandria*, order *Monogynia*, natural order *Stellatæ*, Lin. *Caprifolia*. Juss.

The Cornus florida is of slow growth, and possesses a very compact wood, covered with a rough, broken bark. The branches are smooth, covered with a reddish bark, marked with rings at the place of the former leaves. The leaves, which are small at the flowering time, are opposite, petioled, oval, acute, entire, nearly smooth, paler be-

were about twelve feet high, spreading horizontally, their limbs, meeting and interlocking with each other, formed one vast, shady, cool grove, so dense and humid as to exclude the sun-beams, and prevent the intrusion of almost every other vegetable, affording us a most desirable shelter from the fervid sun-beams at noon day. This admirable grove has by way of eminence acquired the name of Dog woods.

During a progress of near seventy miles through this high forest, there constantly presented to view, on one hand or the other, spacious groves of this fine flowering tree, which must in the spring season, when covered with blossoms, present a most pleasing spectacle, when at the same time a variety of other sweet shrubs display their beauty; as the Halesia, Stewartia, Æsculus, Azalea, &c. entangled with garlands of Bignonia, Glycine, Lonicera, &c. &c. at the same time the superb Magnolia grandiflora standing in front of the dark groves, towering far above the common level." *Travels*, p. 399.

neath, and marked, as in others of the genus, with strong parallel veins. The flowers, which are very small, grow in heads or sessile umbels, upon peduncles an inch or more in length. At the base of each bunch is the large spreading involucrum, constituting the chief beauty of the tree when in flower. This involucrum is composed of four white, nerved, obovate leaves, having their point turned abruptly down or up, so as to give them an obcordate appearance. This point has frequently a reddish tinge. Calyx superior, somewhat bell-shaped, ending in four obtuse spreading teeth. Petals four, oblong, obtuse, reflexed. Stamens four, erect, the anthers oblong with the filaments inserted in their middle. Style erect, shorter than the stamens, with an obtuse stigma. The fruit is an oval drupe of a glossy scarlet colour, containing a nucleus with two cells and two seeds.

The bark of the Cornus florida is a powerful bitter, possessing also an astringent and somewhat aromatic taste. Both tannin and the gallic acid are abundantly developed in its solutions by their proper tests. In my experiments with the bark of young twigs, but a small quantity of pure resin was made manifest. It would seem that the principal seat of the bitterness is in a variety of extractive matter.

In a valuable inaugural dissertation on the Cornus florida and Cornus sericea by Dr. Walker of Virginia, much attention appears to have been bestowed on the chemical properties of their bark. He found that water distilled from the bark in powder had a transparent, whitish appearance, with a slight aromatic odour, and no perceptible taste. When the heat was increased, the fluid had a lemon colour, with an unpleasant smell and an acerb taste. These effects were probably produced by the volatilization and partial decomposition of portions of the bark in consequence of the heat being continued until the mixture was evaporated nearly to dryness.

With a view to ascertain the effect of different menstrua, Dr. Walker subjected to experiment the residual mass furnished by evaporating a decoction of the root of Cornus florida. Two drachms of this residuum, which had been furnished by seven and an half ounces of the decoction, were macerated in successive quantities of the best alcohol, until the last portion ceased to be changed in colour and taste. The part, which remained undissolved, weighed only half a drachm. When redissolved it was destitute of taste, and underwent no change of colour on adding the test of iron. The alcohol, which had been employed in the experiment, was found to possess an intensely bitter taste with astringency, of a clear red colour, and turning to a deep black on the addition of iron. On evaporation, it yielded a drachm and an half of residuum.—Dr. Walker attempted to ascertain the quantity of resin by macerating the alcoholic extract in repeated portions of sulphuric ether. The ether acquired a dark colour and a bitter taste, and was found to have dissolved three quarters of the extract. When tested with iron, it was found that the remaining quarter only was changed to a black colour.

The Cornus florida is one of the many vegetables which, by the union of their gallic acid with the salts of iron, form a black compound, applicable to the purposes of ink. The constancy of the black colour thus produced varies greatly, according to the substance from which the gallic acid is derived. It is often extremely fugacious, sometimes fading in a few days, and at others becoming indistinct after some weeks or months. Considering the very great importance of the purposes for which ink is employed, and the immense evils which may result from its obliteration in writings intended for permanency; it is with extreme caution that we should recommend the introduction of any change in the mode of its formation. The oak gall has had the experience of ages in favour of its permanence and immutability. It is not until some indigenous article, producing an equal intensity of colour, has undergone a series of trials from time and exposure, sufficient to establish beyond a doubt its durability, that its substitution in the manufacture of ink should be considered expedient or even justifiable.

Upon the human body the bark of the Cornus florida acts as a tonic, an astringent and an antiseptic, approaching in its general effects to the character of the Peruvian bark. From a variety of experiments made by Dr. Walker upon the healthy system, it was found that this medicine uniformly increased the force and frequency of the pulse, and augmented the heat of the body. Collateral experiments were made at the same time with the Peruvian bark, with which the Cornus appeared to agree both in its internal and external effects.

In disease it has been principally employed in the same cases for which the cinchona is resorted to, particularly intermittent and remittent fever. Dr. Gregg of Pennsylvania, cited by Dr. Walker, states, that after employing the Cornus florida habitually for twenty three years in the treatment of intermittents, he was satisfied that it was not inferior to the Peruvian bark as a means of cure in such cases. Among the number of cures by this medicine, was that of his own case. He observed that in its recent state it sometimes disagreed with the stomach and bowels, but that this tendency in the article was corrected by age. He recommends the bark as being in the best state after it has been dried a year.

Other medical men have employed the bark of this tree with advantage in intermittents, and also in continued fevers of the typhoid type. Its tonic operation in these cases appears very analogous to that of the Peruvian bark.

I have employed the tincture of Cornus florida as a stomachic in various instances of loss of appetite and indigestion. The report of those who have taken it has perhaps been as frequently in favour of its effects, as of gentian, columbo, and the other imported tonics of the shops, though perhaps it is somewhat more liable to offend the stomach in large doses. In the Southern States a decoction of the buds and twigs has been thought to agree better with weak stomachs, than the other preparations.

Some other species of this family resemble the present tree in the bitterness and tonic power of their bark, particularly the Cornus *circinata* and C. *sericea*.

The wood of the Cornus florida is hard, heavy and fine grained, and susceptible of a good polish. It is employed for various purposes where strength and solidity are required, although its small size does not permit it to be used for objects of much magnitude. From its hardness it is found peculiarly useful for handles of instruments, the teeth of wheels, and the smaller parts of wooden machinery.

BOTANICAL REFERENCES.

Cornus florida, LIN. *Sp. pl.*—GRONOVIUS, *Virg.* 17.—KALM, travels, ii. 321.—WANGENHEIM, *Amer. p.* 51, *t.* 17.—L'HERITIER, *Corn. n.* 3.—SCHMIDT, *Arb. t.* 62—*Botanical Mag. t.* 526.—PURSH, i. 108.—MICHAUX, FIL. *Arbres forestiers*, iii. 138, *translated*, i. 255.—ELLIOTT, *Car.* i. 207.—Cornus mas Virginiana, &c.—PLUKENET, *Alm.* 120, *t.* 2, *f.* 3.—CATESBY, *Car. t.* 27.

MEDICAL REFERENCES.

WALKER, *Inaugural Dissertation, Philad.* 1803.—BART. *Coll.* 12.—THACHER, *Disp.* 203.—ELLIOTT, *ut supra.*

PLATE XXVIII.

Fig. 1. *Cornus florida, a branch with flowers, and one with leaves.*
Fig. 2. *Fruit.*
Fig. 3. *A flower.*

PANAX QUINQUEFOLIUM.

Ginseng.

PLATE XXIX.

In the early part of the eighteenth century some accounts were sent to Europe by travellers and missionaries, of a root growing in Chinese Tartary, known by the name of *Ginseng*, upon which a high value was set by the eastern Asiatics, and which was sold in the cities of China at an enormous price. Father Jartoux, a missionary at Pekin, who had an opportunity of witnessing the collection and use of this root, made a drawing of the plant, accompanied with a particular description, and an account of its uses, and the cause of its high estimation and demand among the Chinese. While on a journey among the mountains of Tartary, performed under the sanction of the emperor of China, he met in various instances

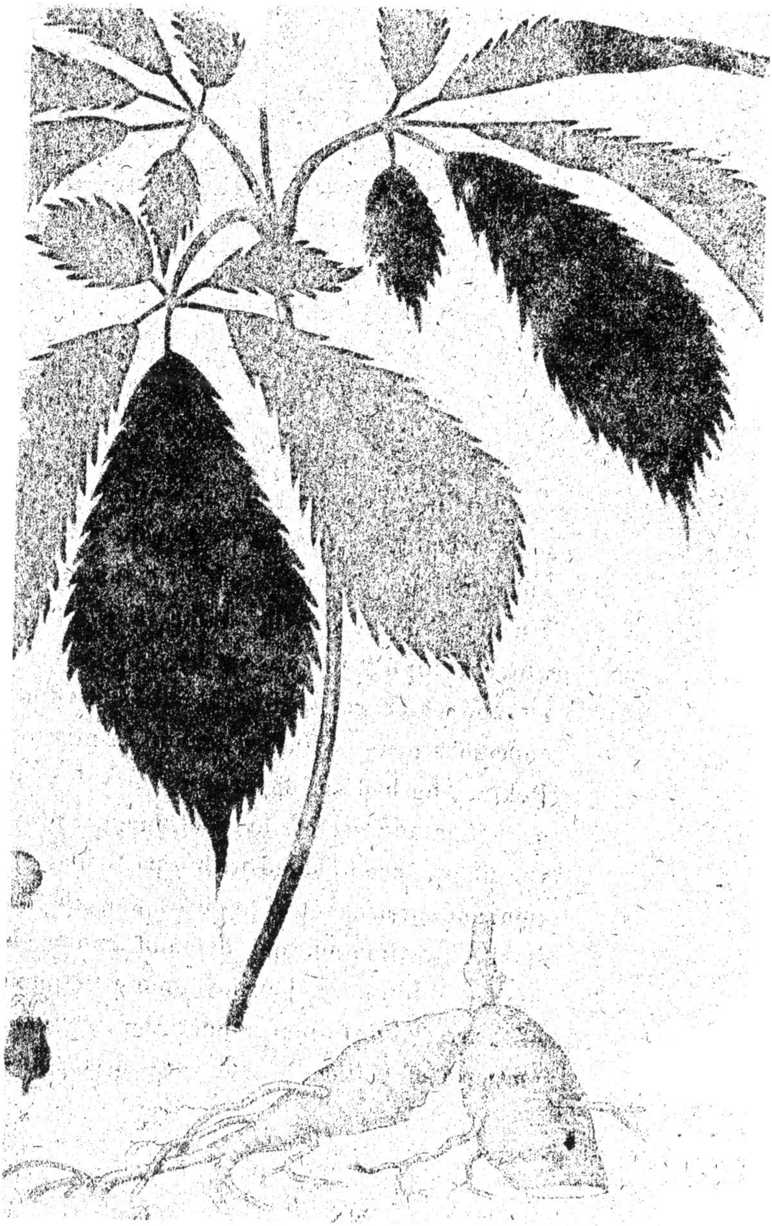

with the plant, and with people employed in collecting it. He states that the root is found principally between the 39th and 47th degree of north latitude, in thick forests, upon the declivities of mountains, on the banks of torrents, and about the roots of trees. It never grows in the open plains or vallies, but always in dark, shady situations, remote from the sun's rays.

As the right of gathering this root is monoppolized by the emperor of China, the most extensive precautions are taken by him to prevent an encroachment on this privilege. The places where the Ginseng is known to grow are guarded with great vigilance, and a whole province, that of Quantong, bordering on the desert, is surrounded by a barrier of wooden stakes, about which guards continually patrole, to keep the inhabitants within bounds, and prevent them from making excursions into the woods, in search of the prohibited drug. Notwithstanding this vigilance, their eagerness after gain incites the Chinese to wander by stealth in the desert, sometimes to the number of two or three thousand, in search of the root, at the hazard of losing their liberty, and all the fruits of their labour, if they are taken. The emperor employs his own servants for the purpose of collection, and in the year 1709, had ten thousand Tartars engaged in scouring the woods in pursuit of the plant. Each man so employed was obligated to present his majesty two ounces of the best he should collect, and to sell him the rest for its weight in pure silver. At this rate it was computed that the emperor would get in a year, about 20,000 Chinese pounds, which would cost him not above one quarter of its value, at the common rate of selling it.

The collectors of the Ginseng carry with them neither tents nor beds, every one being sufficiently loaded with his provision, which is only parched millet, on which he is obliged to subsist during the whole journey. The mandarins send them from time to time some pieces of beef, with such game as they happen to take, which they eat very greedily, and almost raw. They are accustomed to sleep on the ground, and notwithstanding six months are passed in this way, they continue lusty and in perfect health.

The army of herbalists, in order to scour the country effectually, divide themselves into companies of one hundred each, which proceed forward in direct line, every ten of them keeping at a distance from the rest. In this way they overrun an extensive wilderness in a short space of time.

If any one of the company was wanting, as it often happened, either by having wandered out of the way, or being attacked by wild beasts, the party devoted a day or two to search for him, and then returned to their labour.

The root of the Ginseng is the only part preserved. The collectors bury in the ground every ten or fifteen days all that they have procured. In order to prepare it for use, they dip it in scalding water, and scour it with a brush. The roots are then prepared with the fumes of a species of millet, to give them a yellow colour. The millet is put in a vessel with a little water and boiled over a gentle fire. The roots are placed over the vessel upon transverse pieces of wood, being first covered with a linen cloth or another vessel. When treated in this way they assume upon drying a horny or semi-transparant appearance.

The roots may also be dried in the sun, or by the fire, and retain their qualities perfectly. In this case, however, they have not that yellow colour, which the Chinese so much admire.

The Chinese consider the Ginseng as possessing unequalled medicinal powers, and their physicians have written many volumes upon the qualities of the plant. It is made an ingredient in almost all the remedies which they give to their nobility, its price being too expensive for the common people. The sick take it to recover health, and the healthy to make themselves stronger and more vigorous. They affirm that it removes all fatigue, either of body or mind, dissolves humours, cures pulmonary diseases, strengthens the stomach, increases the vital spirits, and prolongs life to old age. Its price at Pekin, according to travellers, has been eight or nine times its weight in silver, and even more.

Father Jartoux became so far a convert to the virtues of the plant, that he tells us that after having taken half of a root, he found his pulse quicker and fuller, his appetite improved, and his strength increased so as to bear labour better than before. On another occasion, finding himself so fatigued and wearied as to be scarce able to sit on horseback, a mandarin in company perceiving his distress, gave one of the roots. He took half of it, and in an hour was not sensible of any weariness. "I have observed," says he, "that the green leaves, especially the fibrous part of them, when chewed, would produce nearly the same effect. The Tartars often bring us the leaves of Ginseng instead of tea, and I always find myself so well afterwards, that I should readily prefer them before the best tea. Their decoction is of a grateful colour, and when one has taken it twice or thrice, its taste and smell become very pleasant."

The Chinese use a decoction of the root, for which they employ about a fifth part of an ounce at a time. This they boil in a covered vessel with two successive portions of water, in order to extract all its virtue.

The following is the substance of Jartoux's description of the Asiatic plant. The root is whitish, rugged and uneven. The stalk is round, and shaded with red; it terminates in a knot or joint at top, from which proceed four equal branches. Each branch produces five leaves, which are equidistant from each other, and from the ground. The leaves are unusually thin and fine, with their fibres very distinguishable, and a few whitish hairs on the upper side. Their colour is dark green above, and a pale, shining green underneath. All the leaves are serrated or finely indented on the edge.—From the centre of the branches rises a second stalk which is very straight and smooth, and whitish from bottom to top, bearing a bunch of round fruit, of a beautiful red colour, composed of twenty four red berries. The red skin of the berry is thin and smooth, and contains a white pulp. As these berries were double, (for they are sometimes found single,) each of them had two rough stones, separated from each other, of nearly the size and figure of common lentils. The berries were supported on small sprigs, which rose from a common centre like the rays of a sphere. The fruit is not good to eat. The berries are not round but a little flat on each side. When they are double there is a depression or hollow place in the middle where the two parts unite. Each berry has a small beard at top diametrically opposite to the sprig on which it hangs. When the berry is dry there remains only a shrivelled skin, adhering close to the stones, of a dark red, or black colour.

The plant dies away and springs up again every year. The number of years may be known by the number of stalks it has shot forth, of which there always remains a mark or scar on the upper part of the root.

"As to the flower," says he, "not having seen it, I can give no description of it. Some say it is white and very small; others have assured me that the plant has none, and that nobody ever saw it. I rather believe that it is so small and so little remarkable, that none of them ever took notice of it.

"There are some plants, which, besides the bunch of berries, have one or two berries like the former, placed an inch or an inch and an half below the bunch. And when this happens, they say if any one takes notice of the point of compass to which these berries direct, he will not fail to find more of the plant."

The foregoing description of Jartoux is introduced as being a very intelligible description of a plant, in language not the most botanical. The drawing, which accompanies the description, is very satisfactory.

The report of the high value of the Ginseng at Pekin led to an inquiry among Europeans, whether the plant was not to be found in parallel latitudes, in the forests of North America. Father Lafiteau, a Jesuit, missionary among the Iroquois, after much search, found a plant in Canada answering the description, and sent it to France. In 1718, M. Sarrasin published in the Memoirs of the Academy an account of the American Ginseng; which, together with one published by Lafiteau the same year, seemed to put its identity with the Chinese vegetable beyond a doubt.

Soon after this the French commenced the collection of the root in Canada for exportation. For this purpose they employed the Indians, who brought it to the merchants for a certain compensation. At one period the Indians about Quebec and Montreal were so wholly taken up in the search for Ginseng, that their services could not be engaged for any other purpose. The American English engaged in the same traffic, and although the plant is a rare one in the woods, yet very large quantities of the root were collected. In 1748, Kalm tells us the common price of the root at Quebec was from five to six livres a pound. The first shipments to China proved extremely profitable to those concerned, especially to the French. In a short time, however, the amount exported overstocked the market, the Chinese began to think the American Ginseng inferior to the Tartarian, and its value depreciated, so that it ceased to be an object of profitable commerce. Its demand has not materially risen at any subsequent period, although it is still occasionally exported. The Chinese most readily purchase the forked or branching roots; and those exporters have been most successful, who have prepared their Ginseng by clarifying it after the Chinese manner.

The American Ginseng is thinly scattered throughout the mountainous regions of the Northern and Middle States. Kalm informs us, that it is seldom found north of Montreal. Michaux states that it inhabits mountains and rich, shady woods from Canada to Tennessee. I have principally met with this plant in the western parts of Massachusetts, and in Vermont, especially on the sides of the Ascutney mountain. Bartram found it near the mouth of the Delaware.

Linnæus has given to the genus of plants, which includes the Ginseng, the name of *Panax*, a Greek word, intended to express the reputed character of the Chinese *panacea*.

The character of this genus consists in *a simple umbel; corolla five petalled; berry inferior, two or three seeded; plants polygamous.*

The species *quinquefolium* has *three quinate leaves.*

The root of this plant consists of one or more fleshy, oblong and somewhat fusiform portions, of a whitish colour, transversely wrinkled, and terminating in various radicles. Its upper portion is slender and marked with the scars of the former shoots. Stem smooth, round, green, with often a tinge of red, regularly divided at top into three petioles, with a flower-stalk at their centre. Petioles round, smooth, swelling at base. Leaves three, compound, containing five, rarely three or seven leafets. The partial leaf-stalks are given off in a digitate manner, and are smooth, compressed and furrowed above. Leafets oblong, obovate, sharply serrate, acuminate, smooth on both sides, with scattered bristles on the veins above. The flowers, which are small, grow in a simple umbel on a round, slender peduncle, longer than the petioles. The involucrum consists of a multitude of short subulate leafets, interspersed with the flower-stalks. These stalks or rays are so short as to give the appearance of a head, rather than umbel. In the perfect flowers the calyx has five small acute teeth; the corolla five petals, which are oval, reflexed and deciduous. Stamens five, with oblong anthers. Styles two, reflexed, persistent; germ large, inferior, ovate-heart shaped, compressed. The berries are kidney shaped, retuse at both ends, compressed, of a bright scarlet colour, crowned with the calyx and styles, and containing two semi-circular seeds. In most umbels there are flowers with only one style, in which case the berry has a semi-cordate form, as represented in fig. 3. Sometimes there are three styles and three seeds. The outermost flowers ripen first, and their berries often obtain their full size before the central ones are expanded. The middle flowers are frequently abortive.

There are also barren flowers, on separate plants, which botanists describe as having larger petals, and an entire calyx. I have not met with plants of this description in flower.

The foregoing character leaves little doubt that the American plant is precisely the same with the Asiatic, although Loureiro and some others have disputed their identity. The description of Jartoux, which has been given, as well as his drawing of the plant, agrees in every respect, except that his plant had four branches or leaves, instead of three. This is accounted for by supposing he had chosen a luxuriant specimen.—It is somewhat remarkable that the names of the Chinese, and of the North American Indians, should signify the same thing in their respective languages, viz. a resemblance to the *figure of a man.* This resemblance, however, it must be confessed, even in the branching roots, is rather of a humble kind.

The genus Panax was placed by Linnæus in his class *Polygamia*, and by late writers in *Pentandria, Digynia.* The plants of this family were also referred by Linnæus to his natural order *Hederaceæ*, or somewhat heterogeneous assemblage of vegetables ; and by Jussieu to his *Araliæ.* Later botanists have placed them among the *Um-*

belliferous vegetables, from which they differ in their berried fruit. The genus most near to Panax is unquestionably *Aralia*, which differs only in the number of styles, a character extremely variable in the Ginsengs.*

The root of the Ginseng has an agreeable taste, consisting of a mixture of sweet and bitter, with some aromatic pungency. Water, both cold and hot, receives a gummy mucus, which is precipitated by alcohol. The watery extract has the taste and smell of the root in a strong degree. The distilled water gives evidence of a volatile oil, and has the aroma, without the sweetness of the root. The common tests indicate the presence of but little resin, and no tannin.

As far as Ginseng has been tried medicinally in this country, and in Europe, its virtues do not appear, by any means, to justify the high estimation of it by the Chinese. That it is not a very active substance, is proved by the fact, that a whole root may be eaten without inconvenience. Its place in the materia medica is among demulcents. It approaches more nearly to liquorice, than to any other medicine in its taste and exter-

* *Panax trifolium*, a beautiful little plant, with nearly the herb of Anemone nemorosa, has always three styles and a tricoccous berry. P. quinquefolium varies from one to three styles, though the usual number is two.

nal qualities. Its extract forms a very neat preparation, and is by no means unpleasant to the taste. Dr. Fothergill tells us, that "in tedious chronic coughs, incident to people in years, a decoction of it has been of service. It consists of a lubricating mucilage combined with some degree of aromatic warmth."

Ginseng is principally sold by our druggists as a masticatory, many people having acquired an habitual fondness for chewing it. It is certainly one of the most innocent articles for this purpose.

BOTANICAL REFERENCES.

Panax quinquefolium, LIN. *sp. pl.*—MICHAUX, *Flora*, ii. 256. —PURSH, i. 191.—WOODVILLE, *Med. Bot.* i. *t.* 58.—*Botanical Mag. t.* 1023.—Aureliana Canadensis, CATESBY, *Car. Suppl. t.* 16.—BREYNIUS *in Prod. rar. p.* 52.—Araliastrum foliis ternis quinquepartitis, Ginseng sen Ninsin officinarum.—TREW, *Ehr. t.* 6, *f.* 1.

MEDICAL REFERENCES.

BOURDELIN, *Hist. de l'Acad.* 1797.—JARTOUX, *tr. in Phil. Trans.* xxviii. 237.—LAFITEAU, *Memoires concernant la precieuse plante de Ginseng. Paris,* 1718.—SARRASIN, *Hist. Acad.* 1718.—KALM, *travels, tr.* iii. 114.—OSBECK, *China, p.* 145.—HEBERDEN, *Med. Trans.* iii. 34.—FOTHERGILL, *Gent. Mag.* xxiii. 209.—CULLEN, *Mat. Med. Vol.* ii. &c.

PLATE XXIX.

Fig. 1. *Panax quinquefolium.*

Fig. 2. *A flower magnified.*

Fig. 3. *Umbel in flower, the external fruit nearly grown.*

Fig. 4. *Germ, calyx and styles magnified.*

Fig. 5. *Root.*

2
3
4
5
6
1

POLYGALA SENEGA.

Seneca Snake root.

PLATE XXX.

THE Seneca snake root has attracted so general an attention from the medical public, as to have become an article of exportation to Europe, and one which holds a regular place in the druggist stores. The plant which produces it has nothing to boast on the score of elegance, and little to attract attention independent of its medicinal virtues. It grows in most latitudes of the United States, especially in the mountainous tracts. The specimen, from which our drawing was taken, was gathered on the borders of Lake Champlain.

The genus *Polygala* has *a five leaved calyx, two of the leaves wing like, and coloured. Capsule obcordate, two celled, and two valved.*

The species *Senega* has *erect, smooth, simple stems, with alternate, lanceolate leaves, broadest at base. Flowers slightly crested.*

Class *Diadelphia*, order *Octandria;* natural orders *Lomentaceæ*, Linn. *Pediculares*, Juss.

The Polygala senega has a firm, hard, branching perennial root, consisting of a moderately solid wood, and a thick bark. This root sends up a number of annual stems, which are simple, smooth, occasionally tinged with red. The leaves are scattered, nearly or quite sessile, lanceolate, with a subacute point, smooth, paler underneath. Flowers white, in a close terminal spike. The calyx, which in this genus is the most conspicuous part of the flower, consists of five leafets, the two largest of which, or wings, are roundish-ovate, white, and slightly veined. Corolla small, closed, having two obtuse lateral segments, and a short crested extremity. Capsules obcordate, invested by the persistent calyx, compressed, two celled, two valved. Seeds two oblong-obovate, acute at one end, slightly hairy, curved, blackish, with a longitudinal, bifid, white appendage on the concave side. The spike opens gradually, so that the lower flowers are in fruit while the upper ones are in blossom.

The rose coloured variety of this plant, as it has been considered by Michaux, proves to be a distinct species. Some species which I possess from Carolina have branching, pubescent stems, and very long, loose spikes. The flowers are several times larger than those of P. senega.

The root of the Polygala senega has an unpleasant and somewhat acid taste. After chewing, it leaves a sensation of acrimony in the mouth, and still more in the fauces, if it has been swallowed. These properties it communicates fully to water upon boiling. The process of decoction does not appear to dissipate any of its power, since the distilled water is destitute of the taste and smell of the plant. Alcohol dissolves a substance, apparently of the resinous kind, giving a precipitate when water is added. Iron produces little change in solutions of this root, and gelatin occasions no alteration whatever.

Medicinally administered, the Seneca snake root is sudorific and expectorant in small doses, and emetic and cathartic in large ones. Its most usual mode of exhibition is in decoction, which may be made of suitable strength by boiling an ounce of the root in a pint and an half of water, till it is reduced to a pint. This preparation may in most cases be given in doses of a table spoonful and upward without disturbing the stomach.

The first reputation of the Seneca root was one which it divides with a multitude of other plants, that of curing the bite of the Rattlesnake. A reward was given by the legislature of Pennsylvania to Dr. Tennent for the promulgation of this supposed property. When, however, we consider the number of cases of recovery from the bite of this serpent, under every variety of treatment, we cannot avoid the conclusion, that these injuries are not necessarily dangerous, and that spontaneous recoveries are perhaps as frequent as those which are promoted by medicine.

More certain success attends the use of the Seneca in pneumonia and some diseases related to it. In the advanced stages of pneumonic inflammation, after venesection and the other usual remedies have been carried to their proper extent; and the cough still remains dry and painful, while the debility of the patient forbids further depletion; in these cases, I have often found a decoction of the Seneca root to afford very marked relief by promoting expectoration, and relieving the tightness and oppression of the chest. Various medical writers have spoken favourably of its employment in these cases, among whom are Lemery and others, in the Memoirs of the French Academy. It has been found injurious, from its stimulating properties, when given at too early a stage, or during the prevalence of much acute inflammation.

Benefit has been derived in *asthma* from the use of this plant. The following is Dr. Bree's opinion, quoted from his treatise on that disease. "Decoction of seneka is eminently useful in the first species, administered to old people, but in the paroxysm of young persons, I have found it too irritating. This distinction applies to convulsive asthma purely uncomplicated, but the disease is frequently observed in middle aged and elderly persons, to take the character of peripneumonia notha in the winter and spring, and seneka is then the most useful medicine that I have tried. In such cases, it should be united with acetated ammonia, during the febrile state, and as this state gives way, the addition of squill, and camphorated tincture of opium, will be found to promote expectoration, perspiration, and urine in a most powerful manner."

In croup, this medicine was introduced into notice by Dr. Archer of Maryland. He speaks with much confidence of its utility in that disease, particularly in promoting the separation and discharge of the membrane formed in the trachea of patients affected by it. Such a membrane, however, does not exist in all cases of croup. And in the early part of the complaint it may be questioned, how far a medicine, which acts as a stimulant to the fauces and neighbouring organs, is entitled to reliance, in a local inflammation of the trachea. It ought not from such a reliance to exclude more active remedies, especially venesection. Dr. Archer's mode of administering it is to give a tea-spoonful of a strong decoction every hour or half hour, according to the urgency of the symptoms, and during the intervals, a few drops occasionally, to keep up a sensible action of the medicine upon the mouth and throat, until it acts as an emetic or cathartic.

In various forms of dropsy, the Seneca root has been resorted to with advantage, and has received the commendations of Percival, Millman, and some others. Its cathartic and diuretic effects are very considerable, when regularly persevered in, in quantities as large as will set easily on the stomach; and have in various instances effected the dissipation of dropsical swellings.

In chronic rheumatism, this root sometimes does good by its universally stimulant and diaphoretic effects. The following case occurred to me some time since in practice. A man labouring under severe rheumatism was ordered to take at intervals a wine glass full of a strong decoction of the Senega made from an ounce of the root in a pint of water. The patient, from a desire to expedite the cure, thought proper to drink the whole quantity at once. The consequence was the most violent vomiting and purging, which lasted the whole night, accompanied with profuse diaphoresis. The patient, as might have been hoped from the violence of the operation, was radically relieved of his disorder.

In uterine complaints, particularly amenorrhea, the Polygala senega has been found of decided efficacy. Dr. Chapman of Philadelphia is one of the authorities for its use in these cases. It must be given largely, and continued for some time.

The most common mode of exhibition of this root is in decoction, as already mentioned. It is also given in powder in doses of twenty or thirty grains. Dr. Tennent likewise employed a wine of Senega made by digesting four ounces of the root in a pound of wine, of which three spoonfuls were given at a dose.

BOTANICAL REFERENCES.

Polygala senega, Lin. *Sp. pl.* Walter, *Car* 178.—Woodville, ii. *t.* 93.—*Bot. Mag. t.* 1051.—Michaux, ii. 53.—Pursh, ii. 464.—Polygala caule simplici erecto, &c. Gronovius, *Virg.* 80.

MEDICAL REFERENCES.

Tennent, *Diseases of Virginia.*—Lemery, Duhamel, &c. *Hist. de l'Acad.* 1739, 136.—Archer, *Med. & Phys. Journal*, i. 83, 106.—Percival, *Med. Journal*, iv. 67.—Bree *on Asthma*, 258.—Massie, *Inaug. Diss. Philad.* 1803.—Thacher, *Disp.* 319.—*N. Eng. Journal*, vii. 206.

PLATE XXX.

Fig. 1. *Polygala senega.*
Fig. 2. *A flower magnified.*
Fig. 3. *Calyx of the same.*
Fig. 4. *Corolla magnified.*
Fig. 5. *Capsule.*
Fig. 6. *A seed.*

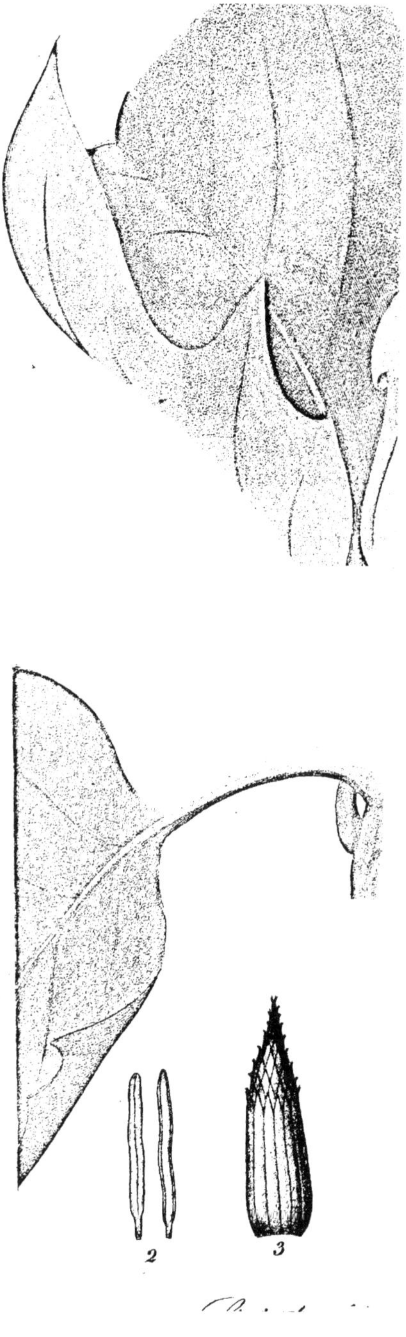
2
3

AMERICAN

MEDICAL BOTANY.

LIRIODENDRON TULIPIFERA.

Tulip Tree.

PLATE XXXI.

THE vegetable world can hardly offer a more interesting object, than a tree of exalted stature and extensive shade, covered with a beautiful and singular foliage, putting forth from its boughs an immense number of large and variegated flowers, at the same time that its trunk affords one of the most useful species of wood, and its bark an aromatic medicinal agent. Such an one is the Tulip tree of the United States.

The forests of the Middle and Western States, according to the representation of Michaux, abound with the Liriodendron tulipifera, as do likewise the elevated parts of Carolina and Georgia. It is found in the New England states, but is principally confined to the southern parts of them.

Cultivated trees are common in Boston and its vicinity, but I have never met with it in the woods of this part of the country, nor to the north of it upon the sea board.

In point of size the Liriodendron is exceeded by few trees of the North American forest. Its growth is regular, straight and majestic. Its trunk often acquires a diameter of from two to three feet, and an elevation of eighty or ninety. In favourable situations it frequently exceeds these dimensions. Michaux measured a tree near Louisville in Kentucky, which at five feet from the ground was twenty two feet and an half in circumference, and which he estimated to be a hundred and twenty or a hundred and forty feet in height. Catesby informs us that the circumference is sometimes thirty feet.

The names of *Tulip tree*, *White wood*, *Canoe wood*, and *Poplar* are applied to this tree in different parts of the United States. Its flowering time is in the months of May and June.

The genus Liriodendron, to which Linnæus has assigned four species of trees, is characterized by *a double calyx, the outer of two, the inner of three leaves; petals six, seeds imbricated into a cone.*

The species tulipifera, the only one in America, is remarkably distinguished by its *lobed and truncated leaves.*

Together with several other of our finest flowering trees and shrubs, the Liriodendron is found in the class *Polyandria* and order *Polygynia*, and the natural orders *Coadunatæ* of Linnæus and *Magnoliæ* of Jussieu.

The branches of the Tulip tree are of a greyish colour inclining to red. The buds which terminate them in winter are very curiously constructed. They are obovate, and flattened or compressed into a sharp edge at the extremity. They are made up of a number of concentric sheaths, each of which contains a single miniature leaf between it and the next interior sheath. This leaf, instead of embracing the next sheath, is folded up and bent down upon one side of it. When vegetation begins in the spring the sheaths swell to a large size before bursting, and at length liberate the leaves one at a time, the remains of each sheath becoming converted into a stipule.

The leaves of the Tulip tree have a form altogether peculiar, and which is not resembled by any other production of our forests. They are divided into four pointed lobes and terminated by a shallow notch, the extremity being nearly square, and the middle rib ending abruptly, as if cut off. In the large leaves, the two lower lobes are furnished with a tooth or additional lobe on their outside. They are attached by long peduncles and have a beautifully smooth and bright green surface. There is one variety of this tree which has the lobes of its leaves not pointed, but very obtuse. The flowers are large, solitary, and terminal. The outer calyx has two triangular leaves which fall off as the flower expands. The inner calyx consists of three large, oval, concave, veined leaves, of a pale green colour, spreading at first, but afterwards reflexed. Petals six, sometimes more, obtuse, concave, veined, of a pale yellowish green, marked with an irregular, indented crescent of bright orange on both sides toward the base. Stamens numerous, with long linear anthers opening outwardly, and short filaments. Pistil a large, conical, acute body, its upper half covered with minute, blackish, recurved stigmas; its lower furrowed, being a mass of coalescing styles and germs. The fruit is a cone of imbricated seed vessels, which are woody and solid, their upper portion formed by a long lanceolate scale. Seeds two, blackish, ovate, one or both often abortive.

The bark of the Tulip tree has a very bitter taste and a strong aromatic pungency. The latter property appears to reside in a volatile oil. When the bark is distilled with water, it fills the apartment with its fragrant odour, yet the product of the distillation, at least when the process is conducted in the small way with the luting of the apparatus not perfectly tight, has scarcely any taste or smell. Dr. Rogers informs us that he obtained an oily matter in the form of a whitish scum on the surface of the water in the receiver. A bitter resin exists in small quantities in the bark. Water dissolves a mucous substance, which is precipitated in a flocculent form by alcohol. Water is also impregnated with the bitterness, and, if too much heat be not employed, with some of the aroma of the tree. The sulphate of iron produced a dark brown colour, but a solution of isinglass did not increase the chemical evidence of astringency, producing a barely perceptible effect. Alcohol and proof spirit may be considered the most perfect solvents of the active ingredients of this article, although water dissolves enough to produce its medicinal effect.

The bark both of the root and branches acts on the system as a stimulating tonic and diaphoretic, having properties resembling the Cascarilla and other aromatic barks of the shops. The disease in which it has been most employed is intermittent fever. But the triumph which results from the occasional cure of this disease is now divided among so large a list of tonic medicines, that the distinction conferred by it is not of the most signal kind. As a warm sudorific, this bark seems well adapted to the treatment of chronic rheumatism, and for this purpose it has been employed with success by various medical practitioners in the United States. In some diseases of an inflammatory type in which it has been recommended, its stimulating properties render it more like to do harm than good. The only personal acquaintance which I have had with it, is as a stomachic. Administered with this view, it has been acceptable and apparently useful to patients who had derived occasional benefit from "Huxham's tincture," "Stoughton's elixir," and similar compositions of bitter and aromatic drugs.

The wood of the Tulip tree is smooth and fine grained, very easily wrought and not liable to split. It is used for various kinds of carving and ornamental work, and for articles of house furniture. In the Western States where pine lumber is scarce, Michaux tells us, that the joinery or inside work of houses is most frequently of this material. A common use of it throughout the United States is in the manufacture of carriages to form the pannels of coach and chaise bodies. For this purpose it is particularly fitted by its smoothness, flexibility and toughness.*

The true or heart wood of this tree is of a yellowish colour and differs in proportion in different trunks. We are told that two varieties of the tree exist, denominated the yellow and the white, and which appear to be in some measure produced by the mode and place of growth. The yellow variety is most valuable, having least alburnum and being less subject than the other to decay.

The Tulip tree has been long since introduced from this country into the forests and fields of Europe. Its use, ornamental appearance, and the facility with which it is raised, have rendered it one of the most prominent and interesting objects of forest cultivation.

* The various economical uses of this tree are treated of at large in the splendid work of the younger Michaux on the *Forest trees of North America.* Those who appreciate the value of a correct knowledge of the various internal resources of our country, will be gratified that a translation of this important work, with the original plates, is now publishing at Paris and Philadelphia.

BOTANICAL REFERENCES.

Liriodendron tulipifera, LINNÆUS, *Sp. pl.*—CURTIS, *Bot. Mag. t.* 275.—MICHAUX, i. 526.—MICHAUX, FIL. *Arbres forestiers,* iii. 202.—PURSH, ii. 382.—Liriodendron foliis angulatis truncatis, TREW, EHRET, 2. t. 10. Tulipifera virginiana &c. CATESBY, *Car.* i. *t.* 48.—PLUKENET, *t.* 117. *f.* 5. &c.

MEDICAL REFERENCES.

RUSH, *Trans. Phil. Col.* i. 183.—BART. *Coll.* 14.—CLAYTON, *Phil. Trans. abr.* viii. 332.—ROGERS, *Inaugural dissertation,* 1802.

PLATE XXXI.

Fig. 1. *A branch of Liriodendron tulipifera.*
Fig. 2. *Stamens presenting different sides.*
Fig. 3. *Pistil.*
Fig. 4. *Fruit.*
Fig. 5. *One of the seeds with its scale or envelope.*

PL. XXXII.

JUGLANS CINEREA.

Butternut.

PLATE XXXII.

Of the forest trees which deserve attention for other properties than the uses of their timber, the Butternut is undoubtedly one of the most interesting; its fruit, bark and juices being all convertible to use. In favourable situations it becomes a large tree, having frequently a trunk of three feet in diameter. It is abundant in the Northern and Middle States, as well as in the Western country. Some parts of the District of Maine, I am told, produce woods of considerable extent, consisting wholly of this tree. Michaux tells us, that it is common in the states of Kentucky and Tennessee, that it abounds on the banks of Lake Erie, the Ohio and even the Missouri. The same author states that it is found in the mountainous parts of Carolina and Georgia, but that he has not met with it in the lower or level portions of the Southern States. It is variously known by the names of *Butternut*, *Oilnut* and *White Walnut*.

The genus *Juglans* or *Walnut* appertains to the Linnæan class *Monœcia* and order *Polyandria*. Its natural orders are *Amentaceæ* of Linnæus and *Terebintaceæ* of Jussieu.

This genus has its *barren flowers in aments with a six-parted calyx; its fertile flowers with a four-cleft superior calyx; a four-parted corolla; two styles; and a coriaceous drupe with a furrowed nut.*

The species *cinerea* has its *leafets numerous, oblong-lanceolate, rounded at base, downy underneath, serrate. Fruit oblong-ovate with a terminal projection, viscid and hairy; nut oblong, acuminate, with a rough, indented and ragged surface.*

The leaves of the Butternut tree when fully grown are very long, consisting of fifteen or seventeen leafets, each of which is two or three inches long, rounded at base, acuminate, finely serrate and downy.

The flowers appear in May before the leaves are expanded to their full size. The barren flowers hang in large aments from the sides of the last year's shoots, near their extremities. The scales which compose them are oblong and deeply cleft on each side into about three teeth or segments. The anthers are about eight or ten in number, oblong and nearly sessile. The fertile flowers grow in a short spike at the end of the new shoot. They are sessile and universally pubescent and viscid. When fully grown, they seem to consist of a large oblong germ and a forked feathery style. The top of the germ, however, presents an obscurely four-toothed calyx. Within this is a corolla of four narrow lanceolate petals growing to the sides of the style. The style divides into two large, diverging, feathery stigmas nearly as long as the germ. These flowers are somewhat later than the aments in their appearance. The fruit is sessile, several together on the sides and extremity of a long peduncle. It is of a green colour, brown when ripe, oblong-oval, pointed, hairy and extremely viscid. It contains a nut which is of a dark colour, carinated on both sides, sharp pointed, its whole surface roughened by deep indentures and sharp prominences. The kernel is more regular than in most nuts of its kind, is very oily, pleasant to the taste when fresh, but acquires a rancid taste by age.

The bark of the branches affords a large quantity of soluble matter, chiefly of the extractive kind. In a concentrated tincture I have not been able to detect any appearance of resin. No evidence of tannin is produced by the test of gelatin. A brownish black colour is caused by the sulphate of iron. The distilled water possesses the taste of the bark in a considerable degree. We are authorized to conclude that water is an adequate solvent for this article, and experience has shewn that the watery extract is one of its best preparations.

The sap of the Butternut tree is saccharine, like that of the Maple, and may be procured in large quantities. In the third volume of the Massachusetts Agricultural Repository is an account of an experiment made on this tree by Mr. M. P. Gray. He states that four trees, the trunks of which were only from eight to ten inches in diameter, produced in one day nine quarts of sap, from which was made one pound and a quarter of sugar. This quantity, it appears from his statement, is equal if not superior to that which the maple affords in the same vicinity.

The inner bark of this tree, especially that obtained from the root, affords one of the most mild and efficacious laxatives which we possess. It is commonly employed in the form of an extract, which preparation is kept in our druggists' shops. Ten or twelve grains of this extract operate gently, and twenty or thirty grains with considerable activity on the bowels. It has been used for many years in this town by the most respectable practitioners. The late Dr. Warren thought highly of its efficacy, and employed it extensively in various complaints, especially in dysentery. During the revolutionary war, when foreign medicines were scarce, this extract was resorted to by many of the army surgeons, as a substitute for more expensive imported drugs. In dysentery it seems at one time to have acquired a sort of specific reputation.

From numerous trials which I have made with this medicine, it appears to me to possess the qualities of an useful and innocent laxative. When fresh and properly prepared, it is very certain in its effect, and leaves the bowels in a good state. In cases of habitual costiveness it is to be preferred to more stimulating cathartics, and many persons whose state of health has rendered them dependent on the use of laxative medicines, have given this the preference after the trial of a variety of other medicines.

A patent medicine, long vended in this state under the name of Chamberlain's Bilious Cordial, was a tincture of this bark combined with various aromatic seeds.

The bark is said to be rubefacient when externally applied, and even capable of exciting a blister. Of this I have had no experience.

BOTANICAL REFERENCES.

Juglans cinerea, LINNÆUS, *Sp. pl.*—JACQUIN, *Ic. rar.* i. *t.* 192.—WILLDENOW, *arb.* 156.—WANGENHEIM, *Amer.* 21. *t.* 9. *f.* 21.—MICHAUX, ii. 191.—PURSH, ii. 636.

Juglans oblonga RETZ. *Obs.* i. p. 10.—Juglans cathartica, MICHAUX, FIL. *Arbres forestiers*, i. 165.

MEDICAL REFERENCES.

THACHER, *Disp.* 245.—BART. *Col.* 23. 32.—RUSH, *Med. Obs.* i. 112.

PLATE XXXII.

Fig. 1. *A branch of Juglans cinerea in flower, the leaves not fully expanded.*

Fig. 2. *A scale or barren flower from the ament magnified.*

Fig. 3. *A fertile flower magnified.*

Fig. 4. *The fruit.*

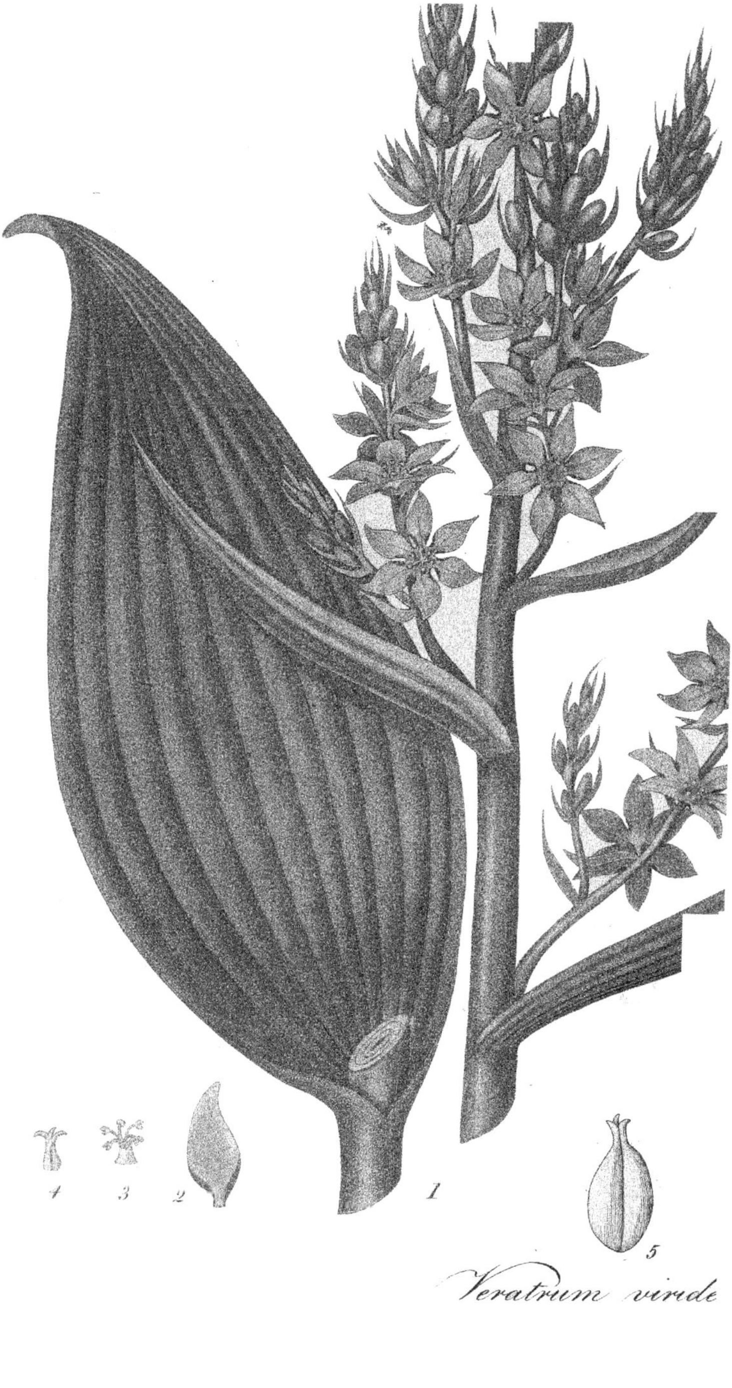

Veratrum viride

VERATRUM VIRIDE.

American Hellebore.

PLATE XXXIII.

In many parts of the United States the swamps and wet meadows, which have been converted into mowing lands, are peculiarly marked in the early part of spring by two species of plants. These are the *Ictodes fœtidus* already described (Pl. xxiv) and *Veratrum viride*, usually denominated *Poke root* and *Hellebore*. Both of these plants spring up more rapidly than the grass around them, and from the largeness and bright green colour of their leaves they are often the most noticeable objects in the places of their growth. As the season advances, the Ictodes continues only a tuft of radical leaves, while the Veratrum sends up a straight leafy stalk, which frequently acquires the full height of a man.

This plant is not only found in boggy meadows, but by the sides of brooks in rocky and mountainous situations, from Canada to Carolina. Its flowering time is from May to July.

The *Veratrum album* or *White Hellebore*, a well known medicinal plant found in most countries of Europe, has a very close resemblance to the American species. It is, however, a smoother plant, and differs somewhat in its flowers, bractes and stalks.

The genus Veratrum, on account of the diversity of its flowers, was placed by Linnæus in his class *Polygamia* and order *Monœcia*. Those more recent botanists, who omit this class, have transferred the genus to *Hexandria trigynia*.

The generic character of *Veratrum* consists in a *six-parted corolla without calyx. Stamens inserted in the receptacle. Capsules three, many seeded. A part of the flowers barren.* The species *viride* has a *downy panicle with the partial bractes longer than their pedicels. Segments of the corolla thickened on the inside at base.*

The root of this plant is thick and fleshy, its upper portion tunicated, its lower half solid and sending forth a multitude of large whitish radicles. The stem is from three to five feet high, roundish, solid, striated and pubescent. Throughout the greater part of its length it is closely invested with the sheathing bases of the leaves. The lower leaves are large, from half a foot to a foot long, oval, acuminate, pubescent, strongly plaited and nerved; the lower part of their edges meeting round the stem. The upper leaves become gradually narrower and the uppermost, which perform the office of bractes, are linear-lanceolate. The flowers are numerous and distributed in compound racemes axillary from the upper leaves, and terminal; the whole forming a sort of panicle. Peduncles roundish, downy. Bractes boat-shaped, acuminate, downy. The pedicel of each flower is many times shorter than its bracte. Calyx none. Corolla divided into six green, oval, acute, nerved segments, of which the alternate ones are longest. All the segments are contracted at base into a sort of claw with a thickened or cartilaginous edge. Stamens six with recurved filaments and roundish, two-lobed anthers. Germs three, cohering, with acute recurved styles as long as the stamens. A part of the flowers are barren and have only the rudiments of styles, so that the plant is strictly polygamous. The seed vessel consists of three capsules united together, separating at top and opening on their inner side. Seeds flat imbricated.

The root of the Veratrum has a bitter taste accompanied with acrimony, and leaves a durable impression on the mouth and fauces when it has been chewed or swallowed. It abounds with a resinous juice, which adheres closely to a knife with which the root has been cut. This resin dissolves abundantly in alcohol. When water is added to the solution, a white turbidness gradually appears, rendering the liquid opaque, but without sediment. The decoction has an intensely bitter taste. It is not rendered turbid by alcohol although some slight flocculi are separated after standing. It is probable that this bitterness resides in an extractive principle. The distilled water of the root has a slightly unpleasant taste, without bitterness or pungency.

This plant in its medicinal powers resembles the Veratrum album or White hellebore of Europe. It is an acrid emetic and a powerful stimulant, followed by sedative effects. As a medicine or as a poisonous plant, it has been known from an early period. The aborigines of the country were fully apprized of its activity. Josselyn in his voyage to New England, which took place not long after the first settlement of the country, informs us that the young Indians had a custom of electing their chiefs by a sort of ordeal instituted with the roots of this plant, which he denominates "white hellebore." A portion of this root was repeatedly given to each individual, and he whose stomach made the most vigorous resistance or soonest recovered from its effects was considered the stoutest of the party and entitled to command the rest.

Kalm tells us that the people of this country, at the time of his travels, employed a decoction of this plant externally in the cure of scorbutic affections, and for the destruction of vermin. He further states, that corn before planting was soaked in a strong decoction of the Veratrum to protect it against the birds which infest our fields and devour the grain after it is deposited in the ground. When the corn is thus prepared, it is observed, that those birds which swallow it become giddy and fall to the ground, an example, the writer informs us, which has the effect to frighten the remainder of the tribe away from the place.

Since the celebrity acquired by the European white hellebore as a remedy for *gout*, that plant being for a time supposed the basis of the celebrated *Eau medicinale;* the attention of some practitioners has been turned to investigating the properties of the American plant, which so closely resembles the Veratrum album in its external habitudes. The result of such trials as have been made, establishes beyound a doubt the medicinal similarity of these two vegetables. I have employed the American plant in dispensary practice in the treatment of obstinate cases of chronic rheumatism. Other practitioners have applied it to the treatment of gout, and of cutaneous and other affections. From the sum of my observations and knowledge respecting it, I am satisfied that the root, when not impaired by long exposure and age, is in sufficient doses a strong emetic, commencing its operation tardily, but continuing in many instances for a long time; in large doses affecting the functions of the brain and nervous system in a powerful manner, producing giddiness, impaired vision, prostration of strength and diminution of the vital powers. Like the Veratrum album and Colchicum autumnale, the violent impression which it makes upon the system has arrested the paroxysms of gout and given relief in some unyielding cases of protracted rheumatism. Like those substances, it requires to be given with great caution and under vigilant restrictions. The solutions of this article have appeared to me more powerful in proportion to their quantity than the substance, probably in consequence of a part of the powder being thrown out in the first efforts to vomit, before a perfect solution of its active parts in the stomach could have taken place.

A course of experiments with this article was made sometime since in the Boston Almshouse by Dr. John Ware, the results of which he has obligingly communicated to me. They cannot be better stated than in his own words.

"I gave this plant," says he, "in the first place with a view to ascertain its action on the stomach and alimentary canal. The doses in which it was administered amounted to from two to ten grains. I began with a small quantity, and increased it very gradually in order to guard against the occurrence of those violent and dangerous effects which I had been led to apprehend from the descriptions given of the operation of the white hellebore. A slight and general account of the experiments will give the most satisfactory view of the effects of this root as an emetic.

"It was administered in about thirty cases. *In the first case* two grains were given; this only produced slight and temporary nausea.

"*In three* instances three grains were administered; in two of these vomiting was produced; in one of them to a considerable degree—in the other slight—in the third no effect whatever was produced.—*Of gr.* iv. Four doses, of which only one operated, and then the operation was inconsiderable.—*Of gr.* vi. Fifteen doses were given—ten of these operated perfectly well; as complete and thorough vomiting was produced as follows from the case of any other emetics—in the eleventh case nausea only ensued—and in the remaining, no effect whatever was perceived.—*Of gr.* viii. Four doses—of these, two failed entirely and two operated satisfactorily.—*Of. gr.* x. Only one dose was given—this operated very thoroughly.

"I did not find, as I had expected, that this substance was uncommonly violent or distressing in its operation. Patients, in general, did not complain of any thing unusual, and when they were particularly questioned as to their sensations, they told of nothing more than those usually occurring during the effects of a brisk emetic. It seemed to produce vomiting rather more severely than an ordinary dose of ipecac—but not more than one of antimony. Indeed, its operation may fairly be said to be about as violent and distressing as that of any other emetic whose effects in evacuating the stomach are equally thorough. In a few instances, however, there was a complaint of very violent and painful retching—

and of dizziness at the time and for a short time after—still these effects were not common nor excessive.

"As to its influence as an emetic upon diseased states of the system, there were few opportunities of administering it where any considerable derangement existed. In those cases which did occur it did not appear to be inferior to the common emetics.

"The degree of operation did not seem to be much increased by the increase of the dose of the medicine. Doses of six grains appeared, when they took effect, to produce vomiting as thorough and complete, as that which followed from larger doses; except that the larger were perhaps more speedy in operating. I could find no cause for the failure of so many of those cases in which the dose amounted to six or eight grains, except an insensibility in the patient to the stimulus of the medicine; and this was rendered more probable from the circumstance that generally in those instances, the substance failed in producing any effect whatever; nausea did not often occur when vomiting was not to follow it, and in no instance was it very clear that purging was produced.

"Indeed this appeared to be rather a singular circumstance relating to this substance, and one in which it differs from most or all other emetics. These articles, when they fail of producing vomiting, generally occasion a determination downwards, and thus produce all the phenomena of cathartic medicines. This effect is also frequently produced when they have operated in their peculiar way. But in no instance did this appear to be the case with the Hellebore. Some patients, indeed, said that it operated upon them by stool very slightly—but on strict inquiry I did not think that the medicine had had any effect in this way, and that what was told me proceeded from a desire in the persons to attribute some sort of effect to what had been given them.

"In the greater number of the cases, the Hellebore was longer before it produced vomiting than is the case generally with other emetics. It did not often operate in less than three quarters of an hour or an hour—but sometimes the interval was extended to two or three hours—and in one case, although the dose was administered at twelve at noon, its effects were not produced until between nine and ten o'clock in the evening. This tardiness in manifesting its effects on the system corresponds with what has been observed with respect to the European species.

"I made a few experiments with the powder of the Veratrum album in order to compare its powers with those which existed in our native species. It was given in six cases—of two doses of three grains—neither produced any effect—one of four was not more effectual—six grains produced some vomiting, but not to any considerable extent—seven no effect whatever—and eight produced in about four hours after taking it considerable effect. If these experiments are to be depended upon, the foreign certainly is not more powerful than the native species.

"I endeavoured next to ascertain what degree of power the Hellebore was possessed of over cutaneous diseases. It has the reputation in the country of some efficacy in these complaints. The ointment and decoction were applied in a number of cases, and it certainly proved to be an application of considerable power. Its effects in some cases amounted to a removal of the disease entirely—and in most, some diminution of it followed, which was more or less durable.

"In one case where there was an eruption about the wrists and other parts of the body, supposed to be the itch—the ointment (made by simmering together a dram of Hellebore in an ounce of lard) after being applied for some time, removed the disease. The cure, however, was protracted and tedious.

"A second case was that of a boy who had on the back of his head, what appeared to be the commencement of Tinea capitis. The ointment made in the same way as that before mentioned, except that simmering was omitted, removed it in the course of a fortnight. It returned again in the course of a few weeks and was again destroyed by the same application. It did not recur so long as I had an opportunity of observing the child, which was, however, not long.

"In one clear case of itch, the ointment failed of producing any beneficial effect, and it was necessary to have recourse to the usual remedy.

"An instance of a very troublesome and irritating eruption upon the hands and fingers, accompanied by a venereal taint of the system, was very much relieved by this ointment. Its final removal, however, could not be attributed to this, since the patient was under the influence of mercury at the same time.

"In some other cases both the ointment and decoction were used with success for a time, but the effects they produced were only partial and temporary."

Of the power of this plant to relieve the formidable disease of gout, we have the most satisfactory evidence. A composition intended to imitate the celebrated *Eau medicinale* was prepared in England by Mr. Moore from the wine of white hellebore and wine of opium in the proportion of three parts of the former to one of the latter. This compound was used by many arthritic patients both in Europe and America with great success in relieving the paroxysms of the disease. In Boston a considerable number of individuals have been induced to make trial of the remedy, and generally with advantage. But I am informed by several of our most repectable apothecaries, that for a long time, especially during the late war, when the white hellebore could not be obtained from Europe, the American plant was used in the preparation of the medicine upon the supposition of its being the same with the European. Various gouty patients made use of it, and no difference was perceived by them or their physicians in its mode of operation or effect upon the disease. Some were relieved by small doses, which did not even nauseate, such as fifteen or twenty drops, repeated if necessary. Others found a drachm of the mixture necessary; which quantity affected them unpleasantly by vomiting or otherwise. Some experienced such severe effects as to deter them after one trial from a repetition of the experiment. In general the paroxysm of gout was completely suspended by the operation of the medicine.

On the whole, we have sufficient knowledge of the American green hellebore, to feel assured that it is a plant of great activity, closely resembling in its properties the Veratrum album of Europe; and that like that plant it has given relief in the paroxysms of gout and in rheumatism. Whether the original *Eau medicinale* be a preparation of Veratrum, Colchicum, or any other acrid narcotic, it is not of consequence here to decide. These plants, with several others that might be mentioned, are similar in their operation, and probably influence the system when under disease, much in the same way. Some individuals obtain relief from moderate doses, which do not occasion nausea or any very disagreeable effects. Others have not derived benefit except from such quantities as bring on vomiting. Some have experienced very distressing consequences, such as excessive sickness, purging, great prostration of strength, impaired vision, and even total insensibility, where the dose has been imprudently large.

The wine of green hellebore is prepared like that of the white, by infusing for ten days eight ounces of the sliced root, in two pints and an half of Spanish white wine. Before being given, in gout, it is combined with one fourth part its quantity of wine of opium. Of this compound the dose varies from twenty drops to a drachm.

From some observations made by Sir Everard Home respecting the wine of Colchicum autumnale, it is probable that the wine of Veratrum may be less violent in its effects, if freed from the sediment which it deposits by standing.

Of the substance of the root freshly powdered, from two to six grains will be found a sufficient dose. For medicinal use, however, in most cases it is probable that the liquid preparation above described promises more.

The external application of the ointment and decoction sometimes produces the same effect on the stomach as the internal use of the plant. In one instance a patient was nauseated and vomiting brought on by the ointment applied to an ulcer of the leg. I have known similar consequences from a strong decoction in cutaneous affections. Might the topical application of this plant be of any service in gout?

BOTANICAL REFERENCES.

Veratrum viride, AITON, *Kew.* iii. 422.—WILLDENOW, *Sp. pl.* iv. 896.—PURSH, i. 242.—REES' *Cyclopedia, ad. verb.*—Veratrum album, MICHAUX, ii. 249.—Helonias viridis, *Bot. Mag.* 1096.

MEDICAL REFERENCES.

KALM, *travels*, ii. 91.—JOSSELYN, *Voyage to New England*, p. 60.—THACHER, *N. Eng. Journal.*

PLATE XXXII.

Fig. 1. *Leaf and flowers of Veratrum viride.*
Fig. 2. *Petal magnified.*
Fig. 3. *Stamens.*
Fig. 4. *Pistils.*
Fig. 5. *Capsule.*
Fig. 6. *Section of the capsule, showing the seeds.*
Fig. 7. *A seed.*

PL. XXXIV.

Gentiana Catesbæi

GENTIANA CATESBÆI.

Blue Gentian.

PLATE XXXIV.

I am indebted to the late Dr. Macbride, of Charleston, S. C. for my specimens of this medicinal plant, and for a quantity of its root in preservation. Many of the perennial species of Gentian have a great degree of bitterness in the root, and among these the *Gentiana lutea*, or common imported Gentian of the shops, stands preeminent. Of the American species, several bear great resemblance in taste and effect to the European plant. No one, however, which I have examined approaches so near to the officinal root in bitterness, as the species which is the subject of this article. This species was formerly confounded with the *Gentiana saponaria*, a fine autumnal plant, common in the Northern and Mid-

18

dle states. It differs widely, however, from that species in the size of its leaves, the length of its calyx, the open mouth of its corolla and shape of its segments. An imperfect figure of this plant, published by Catesby sixty years ago, has been quoted by subsequent botanists as belonging to G. Saponaria. It has now been very properly distinguished by Walter and Elliott, who have applied to the new species the name of its earliest delineator.

The genus *Gentiana* has *a monopetalous corolla; a capsule one-celled and two-valved, with two longitudinal receptacles.* The species *Catesbæi* has *a rough stem; leaves ovate-lanceolate; segments of the calyx longer than the tube; mouth of the corolla open, its outer segments subacute, inner segments bifid and fimbriate.*

This genus is placed by Linnæus in his class and order *Pentandria, Digynia.* A part of the species, however, depart from the character of this class, and constitute one of those instances in which the general rules of the arrangement are violated, that apparently natural genera may not be divided. A part of the Gentians have five stamens and a five-cleft corolla; others have four stamens and a four-cleft corolla; others have a still different number. Yet so uniform is the structure of the fruit, and so great the apparent affinity of the plants, that botanists have hitherto kept the genus entire, even though the variety of form as well as of number in the calyx, corolla, and stamens might perhaps justify a subdivision.

This genus belongs to the natural order *Rotaceæ* of Linnæus, and to Jussieu's *Gentianæ*.

The Gentiana Catesbæi has a branching and somewhat fleshy root. Stem simple, erect, rough. Leaves opposite, ovate or lanceolate, slightly three-nerved, acute, rough on the margin. Flowers crowded, nearly sessile, axillary and terminal. Segments of the calyx linear-lanceolate, varying in length, exceeding the tube and sometimes more than twice its length. Corolla large, blue, ventricase, plaited; its border ten-eleft, the five outer segments roundish and more or less acute, the five inner bifid and fimbriate. Stamens five, with dilated filaments and sagittate anthers. Germ oblong-lanceolate, compressed, supported by a sort of pedicel. Style none, stigmas two, oblong, reflexed. Capsule oblong, acuminate, one-celled, two-valved.

The dried root of this vegetable has at first a mucilaginous and sweetish taste, which is soon succeeded by an intense bitter, approaching nearly to that of the officinal gentian. This quality appears to reside in a bitter extractive principle, soluble in both alcohol and water. A little resin is developed by the pearly appearance which the tincture assumes on the addition of water. The decoction, however, is nearly equal in bitterness to the tincture, and both these solutions exhibit this property much more powerfully than the root in substance. No astringency appears in this root, and nothing remarkable in the distilled water.

I have found the root of this plant in a variety of instances in which I have used it, to resemble very nearly the imported Gentian in its properties; being but little inferior to it in strength or efficacy. Like that substance it invigorates the stomach and gives relief in complaints arising from indigestion. Dr. Macbride, at whose suggestion I first employed it, entertained a high opinion of its tonic power in cases of debility of the stomach and digestive organs.

In Mr. Elliott's Botany of the Southern States, we are told, that in the form of a decoction it is used with decided advantage in cases of pneumonic, where the fever is nervous, and that it acts as a tonic and sudorific. A tincture of it is esteemed as a remedy in dyspepsia, given in doses of one fourth or half an ounce. It is said to increase the appetite, prevent the acidification of the food, and to enable the stomach to bear and digest articles of diet, which before produced oppression and dejection of spirits.

BOTANICAL REFERENCES.

Gentiana Catesbæi, WALTER, *Flora Carol.* p. 109.—ELLIOTT, *Botany of the Southern States,* i. 340.

MEDICAL REFERENCE.

ELLIOTT, *loc. cit.*

PLATE XXXIV.

Fig. 1. *Gentiana Catesbœi.*

Fig. 2. *A flower with the corolla drawn open to show the inner plicœ.*

Fig. 3. *Stamens and pistil in their natural situation.*

Fig. 4. *Stamens separate.*

Fig. 5. *Pistil.*

LAURUS SASSAFRAS.

Sassafras Tree.

PLATE XXXV.

Almost every section of the United States produces the Sassafras tree. It not only inhabits every latitude from New England to Florida, but we are told it is also found in the forests of Mexico and even of Brazil. Its peculiar foliage and the spicy qualities of its bark render it a prominent object of notice, and it seems to have been one of the earliest trees of the North American continent to attract the attention of Europeans. Its character as an article of medicine was at one time so high, that it commanded an extravagant price, and treatises were written to celebrate its virtues. It still retains a place in the best European Pharmacopæias.

Pl. XXXV.

Laurus Sassafras

Annin & Smith Sc.

The genus of trees and shrubs known by the name of *Laurus* comprises many of the most useful as well as celebrated products of the vegetable world. The Bay tree or Laurel of the ancients, the Cinnamon tree, Cassia tree, Camphor tree, and Avocado pear, are either of them sufficient to give notoriety to the genus to which they belong. This genus has *a calyx of from four to six divisions; nectary of three bisetose glands, or wanting; stamens variable in number; fruit a drupe; flowers often polygamous.* The species *Sassafras* is *polygamous, with leaves entire and lobed.*

The Laurels constitute one of the few genera assigned by Linnæus to his class *Enneandria*, to the first order of which they belong. Jussieu has placed them with his *Lauri*, to which they give name. The propriety with which they have been associated with the Linnæan natural order *Oleraceæ* is of a very questionable nature.

The Sassafras tree, of the United State, arrives, in favourable situations, to a tall stature and large circumference. In the Northern States, it is of smaller size, yet trees are sometimes met with about Boston which attain to nearly the average height of the woods around them, and have trunks a foot in thickness. The bark of the trunk is much cracked and of a greyish colour; the young twigs are of a reddish green. The leaves are remarkable for the variety of their form on the same tree. Those which proceed first from the bud are usually oval and entire; the next have the same form with a lobe on one side; the last and most numerous have regularly three lobes. They grow on petioles, and are very downy when young, but become smoother by age. The flowers grow from the sides of the branches below the leaves, having the scales of the former bud for their floral leaves. They are disposed in short slender racemes of a pale green colour, each flower having six oblong segments. Different trees produce barren and perfect flowers. The barren flowers have nine stamens, six of which are exterior and three interior. The perfect flowers, the kind represented in our plate, have only six stamens, with short filaments and heart-shaped anthers. Nectary none. Germ roundish with a straight, erect style. Fruit an oval drupe of a deep blue colour, supported by a red incrassated pedicel. Only a small number of the trees produce fruit.

The bark of this tree has a fragrant smell and a very agreeable spicy taste. The flavour of the root is most powerful, that of the branches more pleasant. The flavour and odour reside in a volatile oil which is readily obtained from the bark by distillation. It is of a light colour, becoming darker by age, very pungent, and heavier than water, so that it sinks in that fluid when the drops are sufficiently large to overcome the repulsion at the surface. The bark and pith of the young twigs abound with a pure and delicate mucilage. A very small quantity of the pith infused in a glass of water gives to the whole a ropy consistance, like the white of an egg. This mucilage has the uncommon quality that it is not precipitated, coagulated, or rendered turbid by alcohol. It continues in a perfectly transparent state when mixed with that fluid, though it does not unite with it. When evaporated to dryness, it leaves a light coloured, gum-like residuum.

The volatile oil and the mucilage appear to contain all the medicinal virtue of the tree.

The bark and wood of the Sassafras were formerly much celebrated in the cure of various complaints, particularly syphilis, rheumatism and dropsy. Its reputation, however, as a specific in those diseases, particularly the first, has fallen into deserved oblivion, while it is now recognized only with regard to its general properties, which are those of a warm stimulant and diaphoretic.

It is retained by the Dispensatories as an ingredient in several preparations, particularly the compound decoction of guaiacum, formerly called "decoction of the woods;" and the compound decoction of Sarsaparilla, formerly the "Lisbon diet drink." These preparations are useful as sudorifics in rheumatism, some cutaneous diseases, and the sequelæ of syphilis. They derive, however, more of their efficacy from their other ingredients, than from the Sassafras, a principal part of the efficacy of which is dissipated by boiling.

The most proper mode of employing the Sassafras is in the form of its volatile oil, which may be given in very small quantities as an antispasmodic, stimulant and sudorific. It is too acrid to be taken unmixed, and should therefore be dissolved in spirit and mixed with water or syrup.

The mucilage of the pith of this tree is peculiarly mild and lubricating, and has been used with much benefit in dysentery, and in catarrhal, as well as calculous affections. Some eminent surgeons have employed it as a lotion in the most inflammatory stages of ophthalmia, to which its softness renders it extremely well suited.

The wood of the Sassafras tree is of a light texture, but is said to be durable when exposed to the weather. It has been thought capable of repelling insects by its odour, and on this account has been employed for trunks, bedsteads, &c. A property of this kind, however, is wrongly attributed to it, since the wood retains scarcely any odour after a few months drying.

BOTANICAL REFERENCES.

Laurus Sassafras, LINN. *Sp. pl.*—PURSH, i. 277.—NUTTALL, i. 259.—WOODVILLE, iv. *t.* 234.—MICHAUX, FIL. *Arbres forestiers*, iii. 173.—Laurus foliis integris, trilobisve.—TREW, *Ehr. t.* 69, 70.—Cornus mas odorata, &c.—PLUKENET, *Alm.* 120, *t.* 222.—CATESBY, *Car.* i. *t.* 55.

MEDICAL REFERENCES.

MURRAY, *Apparatus*, iv. 535.—KALM, *travels*, ii.—HOFFMAN *Obs. Phys. Chem.* 31.—CULLEN, *Mat. Med.* ii. 200.—CLAYTON, *Phil. Trans. Abr.* viii. 332.—BREMANE, *Sassafrasologia* in 1627.

PLATE XXXV.

Fig. 1. *Laurus Sassafras, a branch with perfect flowers of the natural size.*

Fig. 2. *A stamen magnified.*

Fig. 3. *Pistil, ditto.*

Fig. 4. *Fruit.*

APOCYNUM ANDROSÆMIFOLIUM.

Dog's Bane.

PLATE XXXVI.

This is a branching perennial plant, found from Canada to Carolina about the sides of fences and the borders of woods. It has a peculiarly neat aspect derived from its smoothness, its leafless and coloured stalk, bushy top and delicate flowers. Like the other American species, it is a lactescent plant, with a fibrous bark. It attains its flowering period in June and July.

The genus Apocynum has *a bell-shaped corolla; a nectary of five corpuscles surrounding the germ; anthers adhering to the stigma by the middle; follicles two; seeds with down.*

The present species is *glabrous, its stem erect and branching; cymes lateral and terminal; corolla spreading.*

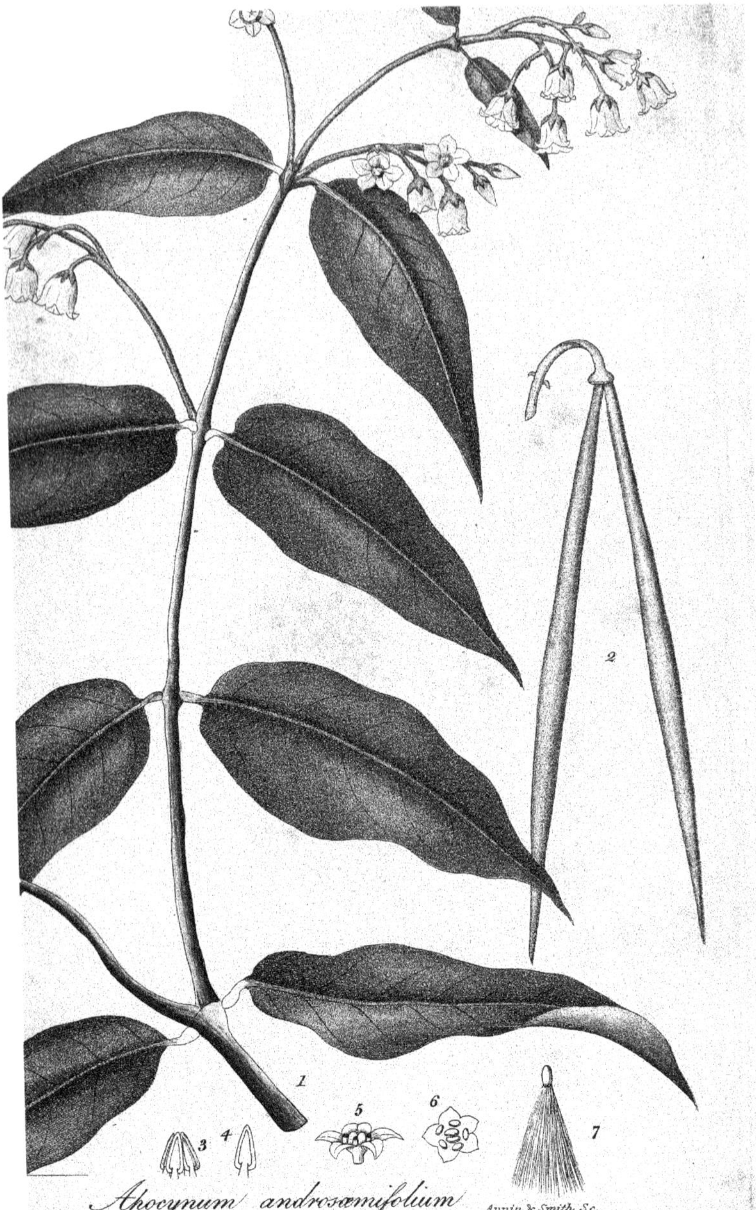

Apocynum androsæmifolium Annin & Smith Sc.

Class *Pentandria*, order *Digynia ;* natural orders *Contortæ*, Linnæus ; *Apocineæ*, Jussieu.

The Apocynum Androsæmifolium grows often to the height of five or six feet, though its common elevation is three or four. Its stalk is smooth, simple below, branching repeatedly at top, red on the side exposed to the sun. Leaves opposite, smooth on both sides, paler beneath, ovate, acute, on short petioles. The flowers grow in nodding cymes from the ends of the branches and axils of the upper leaves, furnished with minute acute bractes. Calyx five-cleft, acute, much shorter than the corolla. Corolla white tinged with red, monopetalous, campanulate, with five acute, spreading segments. Stamens five, with very short filaments, and connivent, oblong arrow-shaped anthers, cohering with the stigma about their middle. The nectary consists of five oblong glandular bodies alternating with the stamens. Germs two, ovate, concealed by the anthers. Stigma thick, roundish, agglutinated to the anthers. The fruit is a pair of slender linear-lanceolate follicles, containing numerous imbricated seeds each crowned with a long pappus or down, and attached to a slender central receptacle.

Every part of the Apocynum when wounded emits copiously a milky juice. When chewed, the root communicates an unpleasant and intensely bitter taste. It exhibits, when dry, the following chemical phænomena.—If a solution in ether be mixed with alcohol, the alcohol, though not turbid at first, becomes so when the ether evaporates. An aqueous infusion or decoction is of a deep red colour and intensely bitter. A solution in alcohol is nearly destitute of colour, but retains the whole bitterness of the plant, and is not disturbed by the addition of water. When submitted to distillation a slight oily film floats on the surface of water in the receiver.—From these facts we may conclude that the Apocynum contains, 1. A bitter extractive principle. 2. A colouring principle soluble in water and not in alcohol. 3. Caoutchouc. 4. A volatile oil.

In various parts of the Eastern States this plant has been shewn to me by country practitioners under the name of *Ipecac.* This name is applied to it from its power of acting on the stomach in the same manner as the Brazilian emetic. Several physicians, among whom is Dr. Richardson of Medway, inform me that they have found about thirty grains of the root to evacuate the contents of the stomach as effectually as two thirds the quantity of Ipecacuanha. In my own trials it has appeared to me much less powerful than the latter substance, and although it produces vomiting, yet this power is diminished by keeping, and appears to be eventually destroyed by age. When used for the purpose of an emetic, the recently powdered root should be employed.

The sensible and chemical qualities of this root seem to promise a good effect when given in small doses as a tonic medicine to the stomach. My observations on this subject may hereafter be more mature. *(See Appendix.)* We have certainly very few indigenous vegetables which exceed the Apocynum in bitterness. Perhaps its emetic property when given in large doses may be owing to this quality. Most bitter vegetables produce vomiting when administered in large quantities.

Kalm observes in his travels in North America, that in some parts of the country this plant was suspected of poisonous properties like those of the Rhus vernix. The country people informed him that the milky juice rubbed on the hands produced blistering in many persons, and that some were affected in the same way even by the effluvium of the plant.—I know of no other authority than that stated above for the existence of such a property in the Apocynum. The plant is common and well known in Massachusetts, yet I have never heard it suspected of deleterious properties.

The flowers of the Apocynum have a power of catching flies and small insects which was ascribed by Dr. Darwin to an irritability in the internal organs. Mr. Curtis in the Botanical magazine, has considered this subject at large, and ascribes the property to a more rational cause. In consequence of the close convergency of the anthers and their adhesion to the stigma, a narrow fissure or slit exists, which becomes more contracted near the top. The insect in search of the honey at the bottom of the flower, inserts his proboscis between the stamens into the cavity within them. In extricating it from this situation the proboscis is sometimes caught in the fissure, and in proportion to the efforts made by the insect to escape it becomes more closely wedged in the upper part of the slit, so that its deliverance by its own powers becomes at length impracticable. Musquitoes, gnats, and small flies are frequently found dead in this confinement.

BOTANICAL REFERENCES.

Apocynum androsæmifolium, LIN. *Sp. pl.*—CURTIS, *Botanical Magazine*, t. 280.—DARWIN, *Botanic Garden*, ii. 182.—MICHAUX, *Flora*, i. 121.—PURSH, i. 179.

MEDICAL REFERENCE.

KALM, *Travels*, iii. 26.

PLATE XXXVI.

Fig. 1. *Branch of Apocynum androsæmifolium.*
Fig. 2. *Pair of follicles.*
Fig. 3. *Cone formed of the united stamens.*
Fig. 4. *Stamen detached.*
Fig. 5. *Side view of the calyx, nectaries and germs.*
Fig. 6. *Front view of the same.*
Fig. 7. *A seed.*

DIRCA PALUSTRIS.

Leather Wood.

PLATE XXXVII.

THE diversity of climate in different latitudes of the United States does not prevent this shrub from appearing in the most rigorously cold as well as in the warmest sections of the country. I have seen it growing plentifully on the banks of the Kennebec, in Maine, and Mr. Bartram found it in great vigour on the Savannah river in Georgia. It is a marshy shrub frequenting low woods and the vicinity of water, flowering in April and May. It is commonly of humble growth, though Mr. Bartram found specimens six or seven feet in height. It is remarkable for the flexibility of its wood and toughness of its bark, which are so great that it cannot be broken without great difficulty. The Aborigines employed

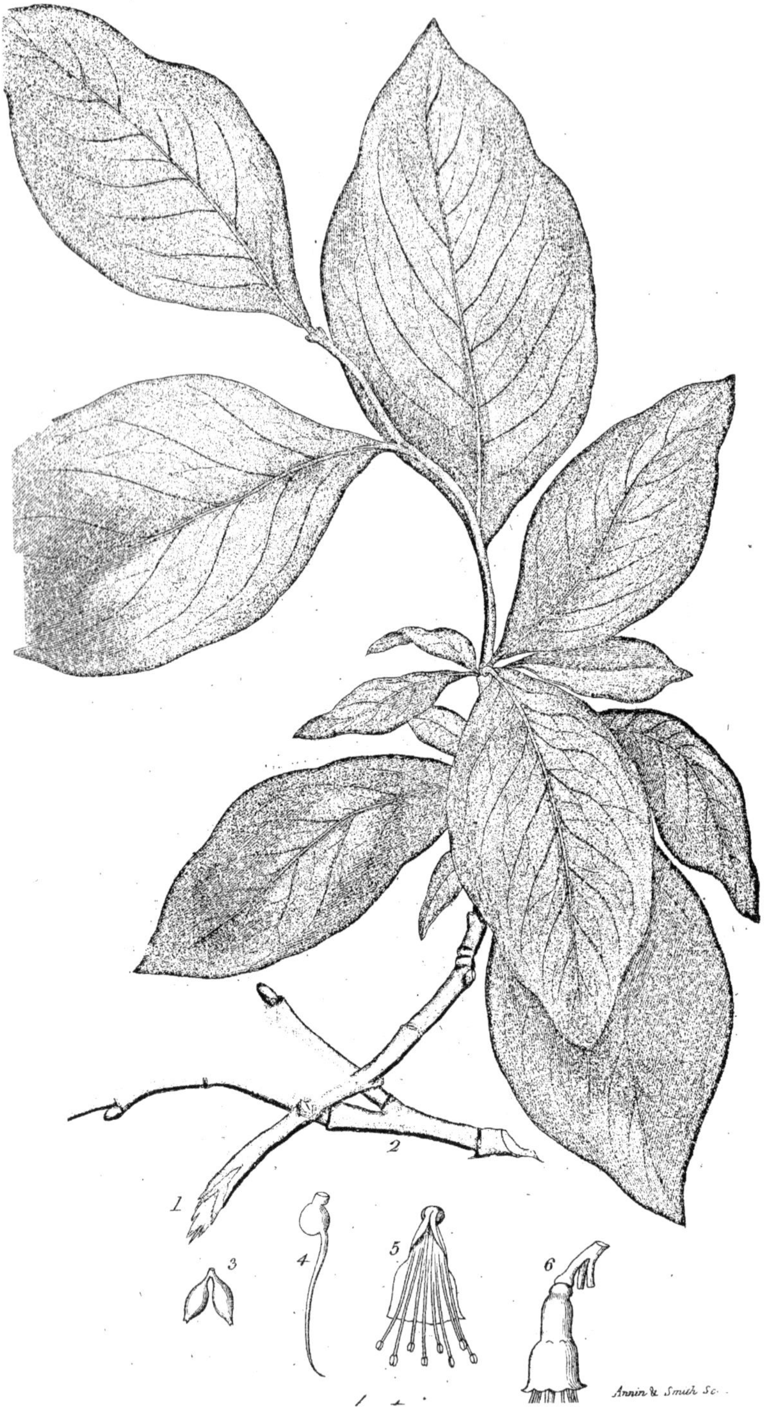
1
2
3
4
5
6
Annin & Smith Sc.

it for their cordage, and from its great tenacity the name of *Leather wood* has been applied to it in most parts of the United States. The Canadian French called it *Bois de Plomb* or *Leaden wood* from its flexibility.

The generic character of Dirca consists in a *tubular corolla without calyx, having its border obsoletely toothed. Stamens longer than the corolla. Berry one-seeded.*

The genus contains but a single species.

Class *Octandria*, order *Monogynia.* Natural orders *Vepreculæ*, Lin. *Thymelææ*, Juss.

The Dirca is an irregular shrub somewhat distinguished by the horizontal tendency of its branches and leaves. The branches have an interrupted or jointed mode of growth. The leaves are scattered or alternate, with very short petioles. They are oval, entire, subacute, downy, when young, smooth and membranous when fully grown, and pale on the under side. The flowers appear long before the leaves. Previously to their emerging they exist in miniature within a small hairy bud, which occupies a sheath or cavity in the end of each flowering branch. They are commonly in bunches of three together with their peduncles cohering. Each flower is about half an inch long, of a yellow colour and without calyx. The corolla is funnel-shaped, with a contraction near the base and another in the middle, its border dilated, and slightly and irregularly toothed. Stamens eight, much longer than the corolla, the alternate ones longest, the filaments capillary and inserted into the tube; the anthers roundish. Germ ovate placed obliquely, the style appearing to issue from one side. The style is capillary, curved, and longer than the stamens. The fruit is a small oval, acute, red, one-seeded berry.

Chemically examined, the bark of this shrub discovers a slightly resinous character by the pearliness which its tincture assumes on admixture with water. The decoction is somewhat mucilaginous and deposits slight flocculi on the addition of alcohol. Iron and gelatin produce no evidence of tannin or gallic acid. The distilled water has an unpleasant odour, but is void of acrimony.

The bark of the Dirca has a peculiar and rather unpleasant taste. When swallowed, it leaves a sensation of acrimony in the fauces which continues for some time. If taken in the quantity of six or eight grains, it produces a sense of heat in the stomach and at length brings on vomiting. This effect pretty certainly occurs if the bark be recent or freshly powdered.

A variety of observations on this shrub have been made by my pupil, Dr. John Locke, who first called my attention to the examination of its properties. He found on experiment that not only the distilled water, but the decoction also was void of acrimony, and that in the boiled bark this property was very much diminished, though still present. The watery extract had considerable bitterness, but scarcely any of the peculiar acrimony of the plant. Taken in doses of a drachm, it did not produce any very sensible effect. Alcohol without heat acquired but slight sensible properties from the bark. Nothing came over by distillation with alcohol, but the alcohol remaining in the retort had acquired the acrimony. The spirituous extract procured by evaporating this decoction was equal to one twenty fourth of the bark from which it was obtained. It contained the acrimony in a concentrated form, producing a more powerful effect on the fauces than the fresh bark. It was largely but not completely soluble in water.

Dr. Locke gave the freshly dried root to various patients in doses of from five to ten grains, which quantity in most instances proved powerfully emetic, and sometimes cathartic. It was found to be deteriorated by keeping, and did not produce the same effects when very old. In consequence of some statements which have been made in regard to its vesicating properties, Dr. Locke applied portions of the bark moistened with vinegar to the skin of his arm. In twelve hours no effect was produced, in twenty four some redness and itching took place and in thirty a complete vesication followed.

The fruit of the Dirca has been suspected of narcotic properties. Dr. Perkins, of Hanover, N. H. has communicated the case of a child which had eaten these berries with effects like those produced by Stramonium, such as stupor, insensibility, and dilatation of the pupils. An emetic brought up the berries and the child gradually recovered. A medical student who took several of the berries found that they produced nausea and giddiness.

The medicinal action of the bark of the Dirca probably depends on its acrid constituents, which appear to be partly of a resinous and partly of a volatile nature. Its properties appear somewhat allied to those of Polygala senega, for which it might perhaps be substituted in small quantities. It is best given in substance, though on account of the tenacity of its fibres it is difficult of subdivision. After beating in a mortar it resembles fine lint more than powder. Its vesicating properties appear too feeble to promise much utility.

I have introduced the Dirca in this place, not so much because it has been yet applied to any medical purpose of great importance; but because it would be improper, in a work like the present, to pass over unnoticed a shrub of such decided activity.

BOTANICAL REFERENCES.

Dirca palustris, LIN. *Amœnitates academicæ*, iii. *t.* I. 6. 7.—DUHAMEL, *Arb.* vi. *t.* 212.—PURSH, i. 236.—MICHAUX, i. 268.—BARTRAM, *Travels*, 309.—KALM, *Travels*, ii. 148.

MEDICAL REFERENCE.

B. S. BARTON, *Coll.* 32.

PLATE XXXVII.

Fig. 1. *A branch with leaves of Dirca palustris.*
Fig. 2. *A branch in flower.*
Fig. 3. *Fruit.*
Fig. 4. *Germ and style much magnified.*
Fig. 5. *Magnified corolla, laid open.*
Fig. 6. *Flower magnified.*

RUBUS VILLOSUS.

Tall Blackberry.

PLATE XXXVIII.

THE family of shrubs comprized under the term *Rubus*, including the various species of *Raspberry*, *Blackberry*, *Dewberry*, &c. is extensively diffused throughout the United States. Many of them are known as troublesome brambles, a few are unarmed, and a certain number are nearly herbaceous. Some are distinguished by the elegance of their flowers, and others by the deliciousness of their fruit. The Rubus villosus is one of the most common and interesting species. It abounds among the brushwood of neglected fields and pastures, about fences and the borders of woods, from the Eastern to the Southern states. Being of rapid growth, it is frequently troublesome to the farmer by spread-

LXXVIII

Rubus villosus

Annin & Smith S

ing in his lands, although it offers some amends for the intrusion by the abundance and fine flavour of its fruit. It is commonly called *tall* or *high* blackberry in distinction from the *R. trivialis* or *low* blackberry, which it greatly resembles in the quality of its fruit. It is in flower in June and its fruit is ripe in August and September.

For the generic character, it has a *five-cleft calyx; five petals; and a compound berry composed of one-seeded acini.*—This species is *pubescent, bristly and prickly, the leaves in threes or fives, leafets ovate, acuminate, serrate, pubescent, with the petioles prickly; flowers racemed.*

Class *Icosandria*, order *Polygynia;* natural orders *Senticosæ*, Lin. *Rosaceæ*, Juss.

This shrub has a tall, branching, prickly stem, which is more or less furrowed and angular. Leaves mostly in threes on a channelled, hairy petiole. A few are solitary and some quinate. Leafets ovate, acuminate, sharply and unequally serrate, covered with scattered hairs above, and with a thick soft pubescence underneath. The terminal leafet is pedicelled, the two side ones sessile. The petiole and back of the middle rib are commonly armed with short recurved prickles. The flowers grow in erect racemes with a hairy, prickly stalk. The pedicels are slender, an inch or two in length, covered with glandular hairs and supported by lanceolate bractes. Calyx divided into five ovate, concave, hairy segments ending in an acuminate point or a lanceolate leafet. Petals five, white, ovate or oblong, concave, contracted into a short claw at base. Stamens very numerous, with roundish anthers and slender, white filaments. Germs numerous, covering a conic central receptacle. Styles capillary, arising from the sides of the germs, persistent. Fruit a black, shining, compound berry formed of pulpy acini attached to the receptacle, each containing a single oblong seed.

The bark of the root of this bramble is the part which has been medicinally employed. It is a pure and strong astringent, which property it manifests both by its sensible effects and by chemical examination. When treated with the sulphate of iron both the tincture and decoction assume a beautiful dark purple colour and throw down a copious precipitate. A precipitate also takes place on the addition of gelatin, which is copious, white and opaque. The alcoholic solution is in part decomposed by water. The substance precipitated does not occasion the uniform turbidness which usually attends the separation of resins, but exhibits a flocculent appearance like that of congulated mucilage. These flocculi, however, when collected and dried, exhibit the common resinous properties on exposure to heat. I subjected the dried bark to distillation, but the distilled water was nearly insipid, possessing only a very slight flavour of the root.

The properties of this bark are those of a very powerful astringent. I have tried its operation sufficiently to become satisfied of its efficacy both internally and externally used in a variety of cases which admit of relief from medicines of its class. It is true that our list of vegetable astringents has become very numerous and the cases which require them are perhaps less frequent than was formerly imagined; yet as we continue to import and consume various foreign medicines of this kind, we ought not to exclude from attention native articles of equal efficacy. Professor Chapman, of Philadelphia, expresses a very decided opinion in regard to the powers of this substance. "Of the vegetable astringents," says he, "this I have reason to believe is among the most active and decidedly efficacious in certain cases. To the declining stages of dysentery after the symptoms of active inflammation are removed, it is well suited, though I have given it, I think, with greater advantage under nearly similar circumstances, in cholera infantum. To check the inordinate evacuations which commonly attend the protracted cases of this disease, no remedy has ever done so much in my hands. Even two or three doses will sometimes so bind the bowels that purgatives became necessary. Being so powerfully astringent, this medicine is useful in all excessive purgings, and especially in the diarrhea of very old people, as well as when it occurs at the close of diseases. During my attendance in our public institutions I had abundant opportunities of testing its efficacy in these cases."

The fruit of the blackberry is among the most delicious productions of the uncultivated forest. To an agreeable combination of sweetness and acid it adds an aromatic fragrance which is surpassed by few of the lighter fruits produced among us. It differs in size and perfection in different seasons, warm and dry summers being most favourable to its perfect maturity. Our markets, however, are rarely destitute of this fine fruit in the months of August and September.

Some other species of Rubus are closely allied to this in the qualities of their fruit and bark, particularly the *Rubus procumbens*, commonly called *low* or *running* blackberry or dewberry.

The fruit of this species is usually larger but produced in smaller quantity from the inflorescence being nearly solitary. The bark is not less astringent than in the present species.

BOTANICAL REFERENCES.

Rubus villosus, AITON, *Kew*, ii. 210.—WILLDENOW, ii. 1085. —MICHAUX, i. 297.—PURSH, i. 346.

MEDICAL REFERENCES.

CHAPMAN, *Therapeutics and Mat. Med.* ii. 474.—THACHER, *Disp.* 341.

PLATE XXXVIII.

Fig. 1. *A specimen of Rubus villosus in flower.*
Fig. 2. *Stamen.*
Fig. 3. *Pistils.*
Fig. 4. *The ripe fruit.*

CASSIA MARILANDICA.

American Senna.

PLATE XXXIX.

This tall and luxuriant plant is found in rich soils in the vicinity of water from New England to Carolina, and westward to the banks of the Missouri. The most northern situation in which I have known it decidedly indigenous, is on the banks of the Quinebaug river near the southern boundary of Massachusetts. It is, however, cultivated in gardens for medicinal use much further to the north. It is a vigorous herbaceous perennial with stalks four or five feet high, having their summits covered in July and August with brilliant yellow flowers.

The extensive genus *Cassia* has *a five-leaved calyx* and *five petals ; anthers unequal, the three uppermost barren, the three lowermost longer,*

t. XXXIX.
6

curved and beaked. Legume two-valved.—The species Marilandica has *eight or nine pairs of leafets, which are oblong-lanceolate, and mucronate; an obovate gland on the petiole. Racemes axillary and terminal; legumes linear and curved.*

Class *Decandria*, order *Monogynia.* Natural orders *Lomentaceæ*, Linn. *Leguminosæ*, Juss.

The stems, which grow in bunches and often attain the height of five or six feet, are round, striated, and invested with a few scattered hairs. Petioles compressed, channelled above, bearing from eight to ten pairs of leafets, which are oblong, smooth, somewhat hairy at the edges, pale on the under side, supported by short crooked pedicels, and mucronated with a rigid bristle at the end. On the base of the petiole is a large ovate pedicelled gland, of a shining green, terminating in a dark point at top, which is sometimes double. Each petiole is also furnished with a pair of linear-subulate, ciliate, deciduous stipules. The flowers grow in axillary racemes, extending quite to the top of the stem. The peduncles are slightly furrowed, pedicels supported by bractes like the stipules, and marked with minute, blackish, glandular hairs. Leaves of the calyx yellow oval, obtuse, the lateral ones longest. Petals five, bright yellow, spatulate, concave, very obtuse, three ascending and two descending. Stamens ten with yellow filaments and brown anthers. The three upper have short abortive anthers. To these succeed two pairs of deflexed linear anthers. The remaining three, or lowermost, are much longer, crooked, and taper into a sort of beak, the middle one being shortest. The anthers open by a terminal pore. Germ descending with the lower stamens, hairy; style ascending, stigma hairy, moist. The fruit consists of long legumes which are pendulous linear, curved, swelling at the seeds, and furnished with slight hairs.

The predominant constituents of the leaves in this plant appear to be resin extractive, and a volatile matter. The tincture is of a dark brown colour and is rendered extremely turbid by water. The infusion and decoction have a lighter colour and the peculiar taste of the plant. The distilled water is nauseous. It is found that both the infusion and decoction answer for medicinal use, yet it is probable that the tincture would be more strongly operative, did not the sedative effects of the alcohol prove a balance for the additional parts of the medicine dissolved.

The Cassia Marilandica is related to the oriental *Senna* in its botanical habit, and nearly resembles it in its medicinal virtues.* Neither of these plants is to be ranked among the most active cathartics, and they require to be taken in much larger quantities than aloes, rhubarb or

* There is no doubt that the true Alexandrian Senna is the product of the *Cassia Senna* of Linnæus and of Willdenow. Lamarck has occasioned an unnecessary confusion on this subject, and misled other botanists, by changing the Linnæan name *C. senna* to *C. lanceolata;* while he has appropriated the name C. senna to the variety β of Linnæus, which is the Italian senna, since very properly named *C. Italica.* See Rees' Cyclopedia, Art. Cassia, &c. The African plant is accounted the most active, although the Italian Senna cultivated in Jamaica, according to Dr. Wright, proved fully equal to it in efficacy.

The greater part of the Senna consumed in the United States is imported from the East Indies. Smaller quantities occasionally reach us from different ports of the Mediterranean and Red seas. The common India senna has a lanceolate leaf narrow and acute; petioles without glands, bearing from five to nine pairs of leaves; and a flat oblong curved legume. Medicinally considered, it is one of the most valuable sorts, operating with mildness and certainty. The facility and cheapness with which it is obtained in India, has long caused it to predominate in our markets.

The India senna, which I have examined, has been very pure, consisting only of leaves of Cassia. The Egyptian has frequently a slight admixture of foreign leaves which are nauseous and bitter.

The Cassia senna would doubtless succeed in our Southern states. The product, consisting of the whole leaves of the plant, must necessarily be large, and would well reward the attention of planters. Ripe seeds may probably be found among the senna of the shops which will vegetate, if not too old. According to Roxburgh and Carey, the Arabian senna cultivated at the Bengal garden is a biennial plant.

jalap, to produce their desired effect. Hence the common form of administering senna is in infusion, a large portion being made to communicate its strength to water at a time. As far as I have been able to observe, about one third more of the Cassia marilandica is required to produce a given effect, than of the C. senna. This objection will prevent it from superseding the senna of the shops, although the facility, with which it may be raised in any part of the United States, will render it a convenient medicine where cheapness is an object. It is already cultivated in gardens for medicinal use, and the infusion and decoction are considerably employed by families and country practitioners.

BOTANICAL REFERENCES.

Cassia Marilandica, Lin. *Sp. pl.*—Martyn, *Cent. t.* 23.—Michaux, *Flora*, i. 261.—Pursh, i. 306.—Nuttall, i. 280.—Cassia mimosæ foliis, &c.—Dillenius, *t.* 260, *f.* 339.

MEDICAL REFERENCES.

B. S. Barton, *Coll.* 32.—Thacher, *Disp.* 178.—Chapman, *Therapeutics.*

PLATE XXXIX.

Fig. 1. *Cassia Marilandica.*
Fig. 2. *The three upper stamens.*
Fig. 3. *The fourth and fifth ditto.*
Fig. 4. *The sixth and seventh ditto.*
Fig. 5. *The three lowest ditto.*
Fig. 6. *A legume.*

Pl. XL.

Nicotiana Tabacum

Annin & Smith Sc.

NICOTIANA TABACUM.

Tobacco.

PLATE XL.

At the time of the discovery of America the Tobacco plant was cultivated by the natives in the West India islands and in different parts of the continent, especially those bordering on the Gulf of Mexico. Whatever may have been its native climate, we need not trace it farther back than this period; and can incur but little risk in considering it as indigenous to the southern parts of the United States in their present enlarged extent. [*Note* B.] It is an annual plant capable of perfecting its flowers and fruit in almost any part of the Union, yet seldom found growing spontaneously except in cultivated grounds or their vicinity.

The genus *Nicotiana* has a *funnel-shaped corolla, with its border somewhat plaited. Stamens inclined; stigma emarginate. Capsule two-celled, two* or *four-valved.* The species *Tabacum*, represented in our plate, has its *leaves ovate-lanceolate, sessile, decurrent; flowers panicled, acute.*

Class *Pentandria*, order *Monogynia.* Natural orders *Luridæ*, Linn. *Solanaceæ*, Juss.

The common Tobacco has a long fibrous root; a stalk five or six feet high, erect, round, hairy, and viscid, branching at top. Leaves sessile; very large, ovate or lanceolate, acuminate, viscid, of a pale green colour. Bractes linear, acute. Flowers forming a panicle on the ends of the stem and branches. Calyx swelling, hairy, glutinous, half as long as the corolla, ending in five acute segments. Corolla funnel-shaped, swelling toward the top, the border expanding, with five acute lobes; the tube of a greenish white, the border red. Filaments inclined to one side, with oblong anthers. Germ ovate, style long and slender, stigma cloven. Capsule ovate, invested with the calyx, two-celled, two-valved, but opening crosswise at top; partition contrary to the valves. Seeds very numerous, small, somewhat reniform, attached to a fleshy receptacle.

It is a remarkable law of the animal economy, that the power of use and habit is capable of reconciling the system to bear with impunity what in its unaccustomed state proves highly deleterious and even fatal. It is a fact that most substances in the Materia Medica lose their effect after the continuance of their use for a certain length of time, so that if we would realize their original operation, we must increase their dose in proportion as the body becomes accustomed and insensible to their stimulus. This is particularly exemplified in the narcotics. Many of these substances, which at first are not only nauseous and disgusting in their sensible qualities, but highly injurious in their influence upon health; are so changed in their effect by habitual use, as to become to those who employ them an innocent and indispensible comfort and a first rate luxury of life.

In its external and sensible properties, there is no plant which has less to recommend it than the common Tobacco. Its taste in the green state is acrid, nauseous and repulsive, and a small quantity taken into the stomach excites violent vomiting, attended with other alarming symptoms. Yet the first person who had courage and patience enough to persevere in its use, until habit had overcome his original disgust, eventually found in it a pleasing sedative, a soother of care, and a material addition to the pleasures of life. Its use, which originated among savages, has spread into every civilized country; it has made its way against the declamations of the learned, and the prohibitions of civil and religious authority, and it now gives rise to an extensive branch of agriculture, or of commerce, in every part of the globe.

Tobacco was in use among the aborigines of America, at the time of its dicovery. They employed it as incense in their sacrificial fires, believing that the odour of it was grateful to their gods. The priests of some tribes swallowed the smoke of this plant to excite in them a spirit of divination, and this they did to a degree which threw them into a stupor of many hours continuance. When recovered from this fit of intoxication, they asserted that they had held a conference with the devil, and had learned from him the course of future events. Their physicians also got inebriated with this smoke, and pretended that while under the influence of this intoxication they were admitted to the council of the gods, who revealed to them the event of diseases. *Harriot.*

In 1559 Tobacco was sent into Spain and Portugal by Hernandez de Toledo, and from thence it was carried into France as a curiosity by Jean Nicot or Nicotius, ambassador at the court of Lisbon, whose name is now immortalized by its application to this genus of plants. From this period the use of tobacco spread rapidly through the continent, and in half a century it was known in most countries in Europe. The rich indulged in it, as a luxury of the highest kind; and the poor gave themselves up to it, as a solace for the miseries of life. Its use became so general and so excessive, that in many countries, the constituted authorities, both of church and state, found it necessary to interpose, and to stop the extravagant indulgence in it by the severest prohibitions. James the First of England, besides writing a book against it, called his "Counterblast to Tobacco," gave orders that no planter in Virginia should cultivate more than one hundred pounds. Pope Urban the Eighth published a decree of excommunication against all who took snuff in the church. Smoking was forbidden in Russia under penalty of having the nose cut off. In Switzerland a tribunal (*Chambre du tabac*) was instituted for the express purpose of trying transgressors in Tobacco. A Turk, who was found smoking in Constantinople, was conducted through the streets of that city with his pipe transfixed through his nose.

Even in this country, where the use of Tobacco originated, we find our puritanic ancestors guarding against its abuse by salutary statutes. In the old Massachusetts colony laws is an act laying a penalty upon any one "who shall smoke tobacco within twenty poles of any house;" or who shall "take tobacco in any inn or common victualling house, except in a private room, so as that neither the master of the said house nor any other guest shall take offence thereat."—In the earliest records of Harvard University soon after its foundation, is a regulation of this kind. "No scholar shall take tobacco, unless permitted by the president, with the consent of their parents and guardians, and on good reason first given by a physician, and then in a sober and private manner."

While the legal authorities in various parts of the world took upon them to control the abuse of this fascinating weed, the literati of different countries entered warmly into the discussion of its merits and its faults. Among its advocates were Castor Duranti and Raphael Thorius, both of whom wrote Latin poems expressly in its praise.

The performance of the latter is entitled a "Hymn to Tobacco," and is very lavish in ascriptions to this plant, which he styles the "gift of heaven and the ornament of earth." So warm were the prejudices of its advocates, that it obtained the reputation of a general panacea, and the catalogue of diseases which it was announced to cure, amounted almost to a complete nosology.

But the opinions of its adversaries were not less extravagant upon the other extreme. It is remarkable that in the days of its first general introduction, no man spoke about it with coolness or indifference, but every one warmly espoused its censure or its praise. Camden, in his life of Queen Elizabeth, says, that men used Tobacco every where, some for wantonness and some for health's sake; and that "with insatiable desire and greediness, they sucked the stinking smoke thereof through an earthen pipe, which they presently blew out again at their nostrils;—so that Englishmen's bodies were so delighted with this plant, that they seemed as it were degenerated into barbarians."

Dr. Venner in a work entitled *Via recta ad vitam longam*, published at London in 1638, gives a brief summary of the injuries done by Tobacco. "It drieth the brain, dimmeth the sight, vitiateth the smell, hurteth the stomach, destroyeth the concoction, disturbeth the humours and spirits; corrupteth the breath, induceth a trembling of the limbs, exsiccateth the winde pipe, lungs and liver; annoyeth the milt, scorcheth the heart and causeth the blood to be adusted. In a word, it overthroweth the spirits, perverteth the understanding, and confoundeth the senses with sudden astonishment and stupiditie of the whole body."

A poetical phillippic, called "Tobacco battered," was published in the reign of King James by Joshua Sylvester, in which he compares Tobacco to gunpowder, and pipes to guns; making the mischief of the two equal. But the most celebrated of all invectives against Tobacco was the "Counterblast" of King James I. That weak monarch gave vent to his prejudices against this herb in a publication, in which he professes to disprove all the alleged grounds for the toleration of Tobacco, and warns his subjects in a most earnest manner not to sin against God, and harm their own persons and goods, and render themselves scorned and contemned by strangers, who should come among them; by persevering in a custom loathsome to the eye, hateful to the nose, and banetul to the brain."

Such were the commotions excited by the introduction and spreading of an article, whose use has now become so common as scarcely to attract notice. This article is the product of several species of Nicotiana, but chiefly of the N. tabacum or Virginian Tobacco, and the N. rustica, sometimes called English Tobacco, and being the sort which Sir Walter Raleigh introduced at the court of Queen Elizabeth. Another species, N. fruticosa, is said to have been cultivated in the East prior to the discovery of America. The Indians on the banks of the Missouri and Columbia rivers cultivate for use the N. quadrivalvis of Pursh and Nuttall. It has been remarked that the Tobacco of warm climates is more mild in its flavour, while that raised in colder latitudes is more strong and pungent. The Bengal Tobacco, of which the sheroots are made, is one of the most weak and mild in its properties. After this is the West India Tobacco which affords the Havanna cigars. Next is the Tobacco of our Southern States, and lastly the Tobacco raised in the northern parts of the Union, which is the most acrimonious and pungent of all.*

* Several varieties of *Nicotiana Tabacum* are cultivated in the United States, of which the principal are the *broad leaved* or *sweet scented*, and the *narrow leaved*.

An elaborate chemical analysis of Tobacco, has been published by M. Vauquelin in the *Annales de Chimie.* His results are, that the broad leaved Tobacco furnishes from its juices the following constituents. 1. A large quantity of animal matter of an albuminous nature. 2. Malate of lime with an excess of acid. 3. Acetic acid. 4. Nitrate and muriate of potash in observable quantities. 5. A red matter soluble in alcohol and water, which swells and boils in the fire, its nature undetermined. 6. Muriate of ammonia. 7. A peculiar acrid, volatile, colourless substance, soluble in water and alcohol, and which appears different from any thing known in the vegetable kingdom. It is this principle which gives to prepared Tobacco its peculiar character, and it is perhaps not to be found in any other species of plant. The medicinal activity of Tobacco evidently resides in this volatile portion, for both the extract and decoction of the plant by long boiling become nearly inert, while the essential or the empyreumatic oil is one of the most deadly poisons known.

Among the substances used by Mr. Brodie in his experiments or vegetable poisons, was the empyreumatic oil of Tobacco prepared by Mr. Brande by distilling the leaves of Tobacco in a heat above that of boiling water. A quantity of watery fluid came over, on the surface of which was a film of unctuous substance, which he calls the empyreumatic oil. Mr. Brodie found that two drops of this oil applied to the tongue of a young cat with an interval of fifteen minutes occasioned death. A single drop suspended in an ounce of water and injected into the rectum of a cat, produced death in about five minutes. One drop suspended in an ounce and a half of mucilage and thrown into the rectum of a dog, produced violent symptoms, and a repetition of the experiment killed him.

Tobacco has been used both as a luxury and prophylactic, and as a medicine. In the former cases it has not been taken internally, but only kept in contact with absorbing surfaces. It is well known, that to the mouth it is applied in substance and in smoke; and to the nose in the form of powder. The opinion which at one time prevailed of its power to prolong life and to secure immunity from diseases is now pretty fully abandoned. It has no prophylactic reputation except as a preservation for the teeth, and in some degree as a protection against the contagion of epidemics. In both these cases it is entitled to a certain degree of confidence, though it is probably inferior to many other substances for both these purposes.

As to its effects upon longevity, the great frequency of its use and the facts and observations of Sir John Sinclair render it improbable that when moderately taken, it has any influence in wearing out the constitution, or abridging the usual period of life. But like all other narcotics its excessive use or abuse must impair the health and engender disease. Of the different modes of using Tobacco, I imagine that smoking is the most injurious, and the most capable of abuse, since in this process the active principles of the Tobacco are volatilized with the smoke, and are extensively applied to the lungs as well as the mouth and nose and fauces.

As a medicine, this plant has been employed in a variety of ways for the alleviation and cure of diseases. Externally it has been applied with benefit in tinea capitis and in complaints occasioned by the presence of insects. In the form of a cataplasm applied to the pit of the stomach it occasions severe vomiting. The prostration of strength and other distressing symptoms which attend this application, must prevent its general employment. Still it may be remembered as an auxiliary in cases where other emetics have failed to operate. A surgeon in the U. S. army informed me that the soldiers had an expedient to exempt themselves from duty, by wearing a piece of tobacco under each armpit, until the most alarming symptoms of real illness appeared in the whole system.

Dr. James Currie has recorded a case of epilepsy cured by the external use of Tobacco. A cataplasm was applied to the stomach for several days about half an hour before the expected return of the paroxysm. A violent impression was produced each time upon the system, the paroxysm prevented and the diseased association effectually broken up. Two cases of obstinate and dangerous intermittent were cured in the same manner by a decoction of half a drachm of Tobacco in four ounces of water, thrown up as an enema, a short period before the time of the paroxysm.

The Tobacco enema was formerly recommended in colic, nephritic complaints, *&c.* Of late years it has been extensively employed in strangulated hernia. In cases of this complaint where the taxis has been ineffectually attempted and the usual auxiliaries have failed, an injection made by infusing half a drachm of Tobacco in eight ounces of boiling water for ten minutes, is found extremely useful. If assisted by the local application of ice to the part, it frequently causes the contents of the sac to return spontaneously, and renders the operation unnecessary, which would be otherwise unavoidable. It operates by its powerfully sedative and relaxing effects, as well as by its cathartic property.

When the infusion is not used, an injection of Tobacco smoke into the rectum frequently produces the same consequences. The smoke may be made to penetrate farther than any liquid, and it is equally efficacious, from the activity of the volatile parts. It was formerly much used in the restoration of persons apparently dead from drowning, but of late years it has gone more into disuse. From the sedative effect of Tobacco, the tendency to syncope and the great prostration of strength which it occasions in ordinary cases; it is probable that its employment in cases of asphyxia from drowning, must assist in extinguishing rather than in rekindling the spark of life.

As a diuretic, Tobacco has been administered internally in doses so small as not to offend the stomach, with very good effect. Dr. Fowler has published a collection of facts relative to its use, principally in dropsy and dysury, from which he concludes it is a safe and efficacious diuretic. In thirty one dropsical cases in which he employed it, eighteen were cured and ten relieved; and out of eighteen cases of dysury, ten were cured and seven relieved. Dr. Ferriar and several subsequent practitioners have found it a valuable diuretic, although Cullen does not speak very encouragingly of its use. At the present day it does not seem to be extensively in use, having passed into neglect rather because more fashionable remedies have superceded it, than because it has really been weighed and found wanting. It will always deserve trial in obstinate dropsical cases (and such cases it must be confessed are not rare) in which the more common remedies have been tried without benefit. Of the various formulas recommended by Dr. Fowler, the Wine of Tobacco is the only one preserved in the Edinburgh and Massachusetts pharmacopœias, being the one which is believed to extract most fully the virtues of the Tobacco. It is made by digesting for a week, an ounce of the dried Tobacco in a pound of Spanish white wine. The dose is from thirty to eighty drops. Dr. Fowler himself however believed the most effectual mode of administering the Tobacco, was in the form of pills of a grain each.

Tobacco has been employed with some success in the *locked jaw*, both of warm and cold climates. Mr. Duncan, surgeon of Grenada, has published in the Edinburgh Journal the account of a very distressing case of this kind, which was relieved and finally cured principally by enemas of Tobacco smoke. These applications generally produced syncope and deathlike sickness in the patient, but by prudent management of them, the disease was entirely overcome, and recovery took place. Dr. Holmes of Worcester county, Mass. exhibited the infusion of Tobacco, to a patient under violent tetanus, after the more common remedies had been fully tried without effect. The spasms were completely removed and the patient recovered.

This powerful medicine has been also employed with some palliative effect in hydrophobia and certain other spasmodic diseases. Its internal use however requires great caution, since patients have in various instances been destroyed by improper quantities administered by the hands of the unskilful or unwary. Notwithstanding the common use and extensive consumption of Tobacco in its various forms, it must unquestionably be ranked among narcotic poisons of the most active class. The great prostration of strength, excessive giddiness, fainting, and violent affections of the alimentary canal, which often attend its internal use, make it proper that so potent a drug should be resorted to by medical men, only in restricted doses and on occasions of magnitude.

BOTANICAL REFERENCES.

Nicotiana tabacum, LIN. *sp. pl.*—AITON, *Kew.* i. 241.—WOODVILLE, *Med. Bot. t.* 77. BLACKWELL, *t.* 146.—PURSH, i. 141.—NUTTALL, i. 132.

MEDICAL AND OTHER REFERENCES.

MURRAY, *Apparatus*, i. 681.—WAFER, *Travels*, 102—HARRIOTT, *Voyage to Virginia.*—HAKLUYT, 75.—EVERARD, *de herba panacea, &c.* 1583.—CHRYSOSTOM MAGNENUS, *Exercitationes* 14, *de Tabaco.*—KING JAMES I. *Works, London*, 1616—SHORT, *Discourses on Tea, Tobacco, &c.*—BIENTEMA, *Tabacologia in* 1690.—HAHN, *Tabacologia, Jenœ.*—GERARD, *Historie of Plants*, 360.—VAUQUELIN, *Annales de Chimie*, 1809.—*Edinburgh Med. Comment.* xl. 327.—DESGRANGES, *Journal de Medicine*, 1791.—CULLEN, *Mat. Med.*—FOWLER, *Med. Reports on Tobacco*, 8*vo, Lond.*—TATHAM, *on the Culture and Commerce of Tobacco, Lond.* 1800.—*Med. and Phys. Journal, Vol.* 24, 25, *et passim.*—DUNCAN, *Repr. in N. Engl. Journal for* 1814.—FERRIAR, *Med. Hist.* i. 75, *and* ii. 152.—POTT. ii. 72, 85, *&c.*—WATTERSTON, *Memoir on the Tobacco plant, Washington*, 1817.

PLATE XL.

Fig. 1. *Nicotiana tabacum.*
Fig. 2. *Capsule.*
Fig. 3. *Ripe capsule opening at top.*
Fig. 4. *Transverse section.*

NOTES.

Note A.

A memoir on the cultivation and use of *Asclepias Syraica*, by J. A. Moller, may be found in Tilloch's Philosophical Magazine, Vol. viii. p. 149. Its chief uses were for beds, cloth, hats and paper. It was found that from eight to nine pounds of the silk occupied a space of from five to six cubic feet, and were sufficient for a bed, coverlet and two pillows.—The shortness of the fibre prevented it from being spun and woven alone. It however was mixed with flax, wool, &c. in certain stuffs to advantage. Hats made with it were very light and soft. The stalks afforded paper in every respect resembling that obtained from rags. The plant is easily propagated by seeds or slips. A plantation containing thirty thousand plants yeilded from six hundred to eight hundred pounds of silk.

Note B.

Tobacco was discovered in Cuba, Florida and Mexico, nearly three centuries ago, and was soon after introduced from this continent into Europe. Whether or not any species of it was cultivated in the East before the discovery of America, is a point of no consequence in regard to its American nativity. The extent of country throughout which it was used by the aborigines of this continent, renders it probable that it must have been cultivated in various parts of America for many centuries previous to its discovery.

The following account of the present mode of cultivating Tobacco in our Southern States is extracted from *An Historical and Practical Essay on the Culture and Commerce of Tobacco.* By William Tatham. London, 1800.

First, of preparing the Tobacco Ground.

"There are two distinct and separate methods of preparing the Tobacco ground: the one is applicable to the preparation of new and uncultivated lands, such as are in a state of nature, and require to be cleared of the heavy timber and other productions with which Providence has stocked them; and the other method is designed to meliorate and revive lands of good foundation, which have been heretofore cultivated, and, in some measure, exhausted by the calls of agriculture and evaporation.

"The process of preparing new lands begins as early in the winter as the housing and managing the antecedent crop will permit, by grubbing the under growth with a mattock; felling the timber with a poll-axe; lopping off the tops, and cutting the bodies into lengths of about eleven feet, which is about the customary length of an American fence rail, in what is called a *worm* or *pannel* fence. During this part of the process the negro women, boys, and weaker labourers, are employed in piling or throwing the brush-wood, roots, and small wood, into heaps to be burned; and after such logs or stocks are selected as are suitable to be malled into rails, make clap-boards, or answer for other more particular occasions of the planter, the remaining logs are rolled into heaps by means of hand-spikes and *skids*; but the Pennsylvania and German farmers, who are more conversant with animal powers than the Virginians, save much of this labour by the use of a pair of horses with a half sledge, or a pair of truck wheels. The burning of this brush-wood, and the log piles, is a business for all hands after working hours; and as nightly revels are peculiar to the African constitution, this part of the labour proves often a very late employment, which affords many scenes of rustic mirth.

"When this process has cleared the land of its various natural incumbrances, (to attain which end is very expensive and laborious,) the next part of the process is that of the hoe; for the plough is an implement which is rarely used in *new* lands when they are either designed for tobacco or meadow.

"There are three kinds of the hoe which are applied to this tillage: the first is what is termed the sprouting hoe, which is a smaller species of mattock that serves to break up any particular hard part of the ground, to grub up any smaller sized grubs which the mattock or grubbing hoe may have omitted, to remove small stones and other partial impediments to the next process.

"The *narrow* or *hilling* hoe follows the operation of the sprouting hoe. It is generally from six to eight inches wide, and ten or twelve in the length of the blade, according to the strength of the person who is to use it; the blade is thin, and by means of a moveable wedge which is driven into the eye of the hoe, it can be set more or less *digging* (as it is termed,) that is, on a greater or less angle with the helve, at pleasure. In this respect there are few instances where the American blacksmith is not employed to alter the eye of an *English*-made hoe before it is fit for use; the industrious and truly useful merchants of Glasgow have paid more minute attention to this circumstance.

"The use of this hoe is to break up the ground and throw it into shape; which is done by chopping the clods until they are sufficiently fine, and then drawing the earth round the foot until it forms a heap round the projected leg of the labourer like a mole hill, and nearly as high as the knee; he then draws out his foot, flattens the top of the hill by a *dab* with the flat part of the hoe, and advances forward to the next hill in the same manner, until the whole piece of ground is prepared. The centre of these hills are in this manner guessed by the eye; and in most instances they approach near to lines of four feet one way; and three feet the other. The planter always endeavours to time this operation so as to tally with the growth of plants, so that he may be certain by this means to pitch his crop within season.

"The third kind of hoe is the *broad* or *weeding* hoe. This is made use of during the cultivation of the crop, to keep it clean from the weeds. It is wide upon the edge, say from ten inches to a foot, or more; of thinner substance than the hilling hoe, not near so deep in the blade, and the eye is formed more bent and shelving than the latter, so that it can be set upon a more acute angle upon the helve at pleasure, by removing the wedge.

Of the Season for Planting.

"The term, *season* for planting, signifies a shower of rain of sufficient quantity to wet the earth to a degree of moisture which may render it safe to draw the young plants from the plant bed, and transplant them into the hills which are prepared for them in the field, as described under the last head; and these seasons generally commence in April, and terminate with what is termed the *long season in May;* which (to make use of an Irishism) very frequently happens in June; and is the opportunity which the planter finds himself necessitated to seize with eagerness for the *pitching* of his crop; a term which comprehends the ultimate opportunity which the spring will afford him for planting a quantity equal to the capacity of the collective power of his labourers when applied in cultivation.

"By the time which these *seasons* approach, nature has so ordered vegetation, that the weather has generally enabled the plants (if duly sheltered from the spring frosts, a circumstance to which a planter should always be attentive in selecting his plant patch) to shoot forward in sufficient strength to bear the vicissitude of transplantation.

"They are supposed to be equal to meet the imposition of this task when the leaves are about the size of a dollar; but this is more generally the minor magnitude of the leaves; and some will be of course about three or four times that medium dimension.

"Thus, when a good shower or season happens at this period of the year, and the field and plants are equally ready for the intended union, the planter hurries to the plant bed, disregarding the teeming element, which is doomed to wet his skin, from the view of a bountiful harvest, and having carefully drawn the largest sizeable plants, he proceeds to the next operation.

Of Planting.

"The office of *planting* the tobacco is performed by two or more persons, in the following manner: The first person bears, suspended upon one arm, a large basket full of the plants which have been just drawn and brought from the plant bed to the field, without waiting for an intermission of the shower, although it should rain ever so heavily; such an opportunity indeed, instead of being shunned, is eagerly sought after, and is considered to be the sure and certain means of laying a good foundation, which cherishes the hope of a bounteous return. The person who bears the basket proceeds thus by rows from hill to hill; and upon each hill he takes care to drop one of his plants. Those who follow make a hole in the centre of each hill with their fingers, and having adjusted the tobacco plant in its natural position, they knead the earth round the root with their hands, until is of a sufficient consistency to sustain the plant against wind and weather. In this condition they leave the field for a few days until the plants shall have formed their radifications; and where any of them shall have casually perished, the ground is followed over again by successive replantings, until the crop is rendered complete.

Of Hoeing the Crop.

"The operation of *hoeing* comprehends two distinct functions, viz. that of hilling, and that of weeding; and there are moreover two stages of hilling. The first hilling commences, as heretofore described, in the preparation of the field previous to planting the crop, and it is performed, as before explained, by means of the peculiar implement called a hilling hoe; the second hilling is performed after the crop is planted, with a view to succour and support the plant as it may happen to want strengthening, by giving a firm and permanent foundation to its root; and it may be effected according to the demand of the respective plants by a dexterity in changing the stroke with the weeding hoe, without any necessity to recur to the more appropriate utensil.

"The more direct use of the weeding hoe commences with the first growth of the tobacco after transplantation, and never ceases until the plant is nearly ripe, and ready to be *laid by*, as they term the last weeding with the hoe; for he who would have a good crop of tobacco, or of maize, must not be sparing of his labour, but must keep the ground constantly stirring during the whole growth of the crop. And it is a rare instance to see the plough introduced as an assistant, unless it be the *flook plough*, for the purpose of introducing a sowing of wheat for the following year, even while the present crop is growing; and this is frequently practised in fields of maize, and sometimes in fields of tobacco, which may be ranked amongst the best fallow crops, as it leaves the ground perfectly clean and naked, permitting neither grass, weed, nor vegetable, to remain standing in the space which it has occupied.

Of Topping the Plant.

"This operation, simply, is that of pinching off with the *thumb nail** the leading stem or sprout of the plant, which would, if left alone, run up to flower and seed; but which, from the more substantial formation of the leaf by the help of the nutritive juices, which are thereby afforded to the lower parts of the plant, and thus absorbed through the ducts and fibres of the leaf, is rendered more weighty, thick, and fit for market. The qualified sense of this term is applicable to certain legal restrictions founded upon long experience, and calculated to compel an amendment in the culture of this staple of the Virginia trade, so that it shall at all times excel in foreign markets, and thus justly merit a superior reputation. I do not exactly recollect the present limitation by law, which has changed, I believe, with the progress of experience; but the custom is to top the plant to nine, seven, or five leaves, as the quality and soil may seem most likely to bear.

Of the Sucker, and Suckering.

"The *sucker* is a superfluous sprout which is wont to make its appearance and shoot forth from the stem or stalk, near to the junction of the leaves with the stem, and about the root of the plant; and if these suckers are permitted to grow, they injure the marketable quality of the tobacco by compelling a division of its nutriment during the act of maturation. The planter is therefore careful to destroy these intruders with the thumb nail, as in the act of *topping*, and this process is termed *suckering*.

"This superfluity of vegetation, like that of the top, has been often the subject of legislative care; and the policy of supporting the good name of the Virginia produce has dictated the

* "Many of the Virginians let the thumb nail grow long, and harden it *in the candle*, for this purpose: not for the use of *gouging* out people's eyes, as some have thought fit to insinuate."

wisdom of penal laws to maintain her good faith against imposition upon strangers who trade with her. It has been customary in former ages to rear an inferior plant from the sucker which projects from the root after the cutting of an early plant; and thus a *second* crop has been often obtained from the same field by one and the same course of culture; and although this scion is of a sufficient quality for smoking, and might become preferred in the weaker kinds of snuff, it has been (I think very properly) thought eligible to prefer a prohibitory law, to a risk of imposition by means of similitude.

"The practice of cultivating *suckers* is on these accounts not only discountenanced as fraudulent, but the constables are strictly enjoined *ex officio* to make diligent search, and to employ the posse comitatus in destroying such crops; a law indeed for which, to the credit of the Virginians, there is seldom occasion; yet some few instances have occurred, within my day, where the constables have very honourably carried it into execution in a manner truly exemplary, and productive of public good.

Of the Worm.

"There are several species of the worm, or rather *grub* genus, which prove injurious to the culture of tobacco; some of these attack the root, and some the leaf of the plant; but that which is most destructive, and consequently creates the most employment, is the *horn* worm, or large green tobacco worm. This appears to me to be the same species with that which Catesby has described in the second volume of his Natural History of Carolina, p. 94, under the title *eruca maxima cornuta*, or the great horned caterpillar.

"'This caterpillar,' says he, 'is about four inches long, besides the head and tail; it consists of ten joints, or rings, of a yellow colour; on the head, which is black, grow four pair of horns, smooth and of a reddish brown towards the bottom, jagged or bearded, and black towards the top; on each of the rings arise short, jagged, black horns, one standing on the back, and two on each side; below which is a *trachæa* on each side; likewise the horn of the back of the last ring is longest: the flap of the tail is of a bright bay colour. It hath eight feet, and six *papillæ*.'

"There are, besides this kind, others without horns; all of them of a green colour, so far as I recollect. And this, in Catesby's description, differs in respect to colour; this tobacco worm or *horn* worm, as the planters call it more particularly, being of a pale delicate green; an effect I apprehend which proceeds from the colour of its food when it feeds upon growing tobacco plants. The act of destroying these worms is termed *worming* the tobacco, which is a very nauseous occupation, and takes up much labour. It is performed by picking every thing of this kind off the respective leaves with the hand, and destroying it with the foot.

Of the Term "Firing."

"During very rainy seasons, and in some kinds of unfavourable soil, the plant is subject to a malady called *firing*. This is a kind of blight occasioned by the moist state of the atmosphere, and the too moist condition of the plant: I do not recollect whether the opposite extreme does not produce an effect something similar. This injury is much dreaded by the planter, as it spots the leaf with a hard brown spot, which perishes, and becomes so far a loss upon the commodity. I apprehend there are two stages when the plant is, in a certain degree, subject to this evil effect: the first is whilst growing in the field, the latter when hanging in the tobacco house. I know of no other remedy than constant working the ground while the seed is growing, and careful drying by the use of fire in the tobacco house.

Of the Ripening of the Crop.

"Much practice is requisite to form a judicious discernment concerning the state and progress of the ripening leaf; yet care must be used to cut up the plant as soon as it is sufficiently ripe to promise a good curable condition, lest the approach of frost should tread upon the heels of the crop-master; for in this case, tobacco will be among the first plants that feel its influence, and the loss to be apprehended in this instance, is not a mere partial damage by nippling, but a total consumption by the destruction of every plant.

"I find it difficult to give to strangers a full idea of the ripening of the leaf: it is a point on which I would not trust my own experience without consulting some able crop-master in the neighbourhood; and I believe this is not an uncustomary precaution among those who plant it. So far as I am able to convey an idea, which I find it easier to understand than to express, I should judge of the ripening of the leaf by its thickening sufficiently; by the change of its colour to a more yellowish green; by a certain mellow appearance, and protusion of the web of the leaf, which I suppose to be occasioned by a contraction of the fibres; and by such other appearances as I might conceive to indicate an ultimate suspension of the vegetative functions.

Of Cutting and Gathering the Crop.

"When the crop is adjudged sufficiently ripe to proceed to cutting, this operation is assigned to the best and most judicious hands who are employed in the culture; and these being provided each with a strong sharp knife, proceed along the respective rows of the field to select such plants as appear to be ripe, leaving others to ripen; those which are cut are sliced off near to the ground, and such plants as have thick stalks or stems are sliced down the middle of the stem in order to admit a more free and equal circulation of air through the parts during the process of curing, and to free the plant, as far as possible, from such partial retention of moisture as might have a tendency to ferment, and damage the staple. The plants are then laid down upon the hill where they grew, with the points of the leaves projecting all the same way, as nearly as possible, so that when the sun has had sufficient effect to render them pliable, they may more easily and uniformly be gathered into *turns* by the gatherers who follow the cutting.

Of Gathering the Crop in.

"For the better comprehending the method of gathering the crop, it is necessary to understand the preparation which must be previously made for facilitating this part of the process.

"In preparing for gathering the crop of tobacco it is customary to erect a kind of scaffold in various places of the tobacco ground which may happen to offer a convenient situation. This is done by lodging one end of several strong poles upon any log or fence which may be convenient, and resting the other end of such poles upon a transverse pole supported by forks, at about five feet from the ground; or by erecting the whole scaffold upon forks if circumstances require it.

"In forming this part of the scaffold in the manner of joists, the poles are placed about four feet asunder from centre to centre, so that when the sticks which sustain the tobacco plants are prepared they may fill the space advantageously by leaving but little spare room upon the scaffold.

"Timber is then split in the manner of laths, into pieces of four feet in length, and about an inch and a half diameter. These are termed the *tobacco sticks;* and their use is to hang the tobacco upon, both by lodging the ends of this stick upon the poles of the scaffold which have been previously prepared in the field, in order to render it sufficiently pliable and in condition to carry into the tobacco-house, to which it is now convey-ed by such means as the planter has in his power; and by suspending it in the same way in the house, so that the air may pass through it in the process of curing. Instead of this particular method, those who prefer to do so, lay it a short while in bulk upon poles, logs, &c. in the field, before they convey it under cover."

CONTENTS OF THE SECOND VOLUME.

AMERICAN

MEDICAL BOTANY,

BEING A COLLECTION

OF THE

NATIVE MEDICINAL PLANTS

OF THE

UNITED STATES,

CONTAINING THEIR

BOTANICAL HISTORY AND CHEMICAL ANALYSIS,

AND PROPERTIES AND USES

IN

MEDICINE, DIET, AND THE ARTS,

WITH

COLOURED ENGRAVINGS.

BY JACOB BIGELOW, M. D.

RUMFORD PROFESSOR, AND PROFESSOR OF MATERIA MEDICA IN
HARVARD UNIVERSITY.

VOL. III.

BOSTON:

PUBLISHED BY CUMMINGS AND HILLIARD, AT THE BOSTON
BOOKSTORE, NO. 1 CORNHILL.

UNIV. PRESS....HILLIARD AND METCALF.
1820.

PREFACE TO THE THIRD VOLUME.

THE subjects of the present work, for reasons which prevail in many publications of the kind, have been inserted without reference to any particular arrangement or system. Those plants received the earliest place, the observations respecting which were earliest matured, and the drawings of which were first completed. Although this plan has been objected to in some foreign criticisms, it is the one pursued in several of the most extensive and useful botanical works of the day, which are accompanied with plates; and in periodical publications, or those which appear in successive numbers, it has more than one decided advantage. It gives time for all the figures to be completed at leisure, from perfect specimens, in proper and convenient seasons; at the same time that it does not necessitate premature and imperfect descriptions of their subjects, which must take place were an arrangement adopted, which might require the first insertion for plants not yet obtained or imperfectly examined. A systematic method may be adhered to in a work which is furnished for the press at once, but must occasion delay and imperfection in a periodical one.

As the American Medical Botany is terminated by the completion of its third volume, the opportunity is now afforded for taking a methodical view of its contents. Considered in a medicinal point of view, the subjects will be best classed as in systems of Materia Medica, by a reference to their leading properties or most striking modes of operating on the human system. In this light they may be arranged as follows.

Narcotics.
Datura Stramonium,
Conium maculatum,
Cicuta maculata,
Hyoscyamus niger,
Nicotiana tabacum,
Solanum dulcamara,
Kalmia latifolia?

Astringents.
Geranium maculatum,
Statice Caroliniana,
Arbutus Uva ursi,
Rubus villosus,
Rhododendron maximum,
Nymphæa odorata,
Myrica cerifera.

Tonics.
Menyanthes trifoliata,
Humulus Lupulus,
Eupatorium perfoliatum,
Coptis trifolia,
Cornus florida,
Gentiana Catesbæi,
Aletris farinosa,
Polygala rubella,
Sabbatia angularis,
Prinos verticillatus,
Liriodendron tulipifera,
Magnolia glauca.

Acrid stimulants.
Arum triphyllum,
Ictodes fœtidus,
Ranunculus bulbosus.

Emetics.
Lobelia inflata,
Phytolacca decandra,
Gillenia trifoliata,
Veratrum viride,
Sanguinaria Canadensis,
Iris versicolor,
Apocynum androsæmifolium,
Dirca palustris,
Euphorbia ipecacuanha,
Euphorbia corollata.
Erythronium Americanum.

Cathartics.
Podophyllum peltatum,
Juglans cinerea,
Triosteum perfoliatum,
Cassia marilandica,

Diaphoretics.
Aristolochia serpentaria,
Asarum Canadense,
Xanthoxylum fraxineum,
Solidago odora,
Gaultheria procumbens,
Laurus sassafras,
Illicium Floridanum.

Diuretics.
Juniperus communis,
Pyrola Umbellata.

Expectorants.
Polygala senega,
Asclepias tuberosa.

Demulcents.
Panax quinqefolium.

Anthelmintics.
Spigelia marilandica.

External stimulants.
Juniperus Virginiana,
Rhus Vernix,
Rhus radicans.

We avail ourselves of classification in the Materia Medica founded on the kind of operation which medicines exert on the human body, because there are seemingly no better characteristics by which to arrange them. But even this method is defective, because few medicines are simple in their operation, and of course most of them have claims to stand in more than one class. As examples, Tobacco, Henbane, Foxglove, and Opium are all of them properly placed by authors under the head of Narcotics. But of these, Tobacco is an emetic, Henbane a cathartic, Foxglove a diuretic, and Opium, while it checks all other excretions, is itself sudorific. Mercury, under its different forms and modes of administration, is capable of fulfilling half a dozen different intentions. The classifier of medicines then can do no more than to arrange them by their most obvious and well known properties, whatever these may be, leaving it understood that the name of a class is by no means fully descriptive of the character of its contents.*

In forming a selection of sixty plants to be represented in this work, it has been endeavoured to choose those which are among the most interesting to botanists, at the same time that they possess claims upon the attention of medical men. It is by no means to be asserted that all these possess so decided an efficacy as to entitle them to the rank of standard medicines, or to make it advisable that pharmacopœias should be swelled by their introduction. A part of them no doubt are eminently entitled to this distinction. Others are efficacious only in a second degree,

* For a botanical arrangement of the plants, see the systematic index at the end of the volume.

but are still in use, and often advantageously so, in the hands of country practitioners. There are some of yet inferior efficacy, which, having formerly enjoyed a certain degree of medicinal notoriety, are inserted here with a view of defining their true character.

The progress of botanical students is much facilitated by the possession of correct drawings and dissections of a variety of dissimilar plants. In this country botanical figures, especially of American plants, are scarce, and accessible to but a small number of those who pursue this study. It is hoped that the present work may, in a certain degree, supply the deficiency, at least until the extension of natural science among us, and the increased number of botanical students, shall call forth and support works of greater magnitude.

A part of the plants contained in this work have never been figured in any botanical work. Others have been represented a great number of times; yet their importance, in a medical point of view, required their admission; and the figure being always made from an American specimen, it may, on this account, be not destitute of interest.

Having arrived at the termination of the American Medical Botany, the author feels it incumbent on him to state, that he has at no time had cause to regret the undertaking of a work, which has furnished a most interesting employment for his leisure hours; and which has been honored with a patronage, greatly exceeding his anticipations.

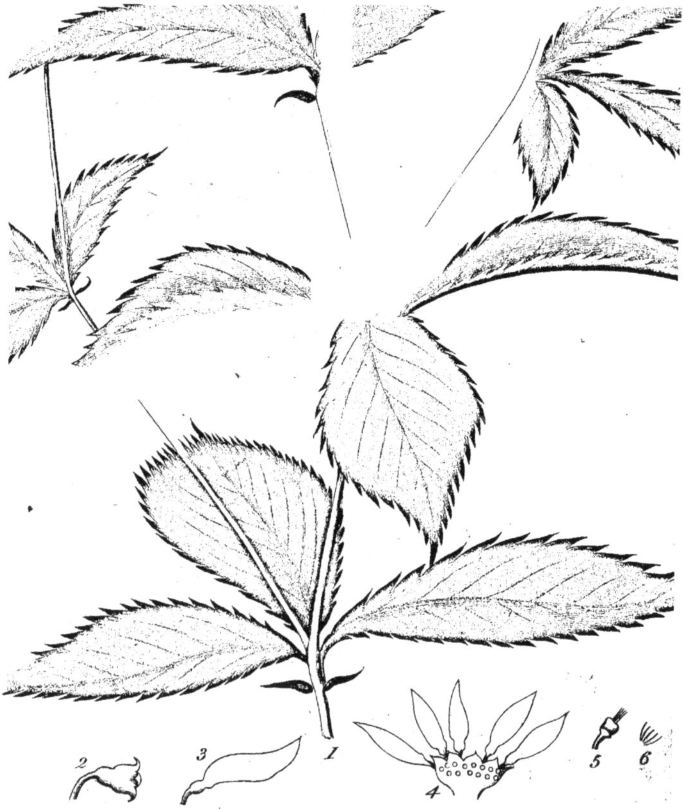

Gillenia trifoliata

AMERICAN

MEDICAL BOTANY.

GILLENIA TRIFOLIATA.

Common Gillenia.

PLATE XLI.

NOTWITHSTANDING the principle avowed by Linnæus, that genera are formed by nature; the determination of generic consanguinity in species occasions in many instances one of the greatest perplexities of the botanist. What difference in structure and external form either of flower or fruit, is sufficient to separate families of plants from each other; is a point often difficult to decide; and is perhaps as frequently set at rest by convenience and by arbitrary decision, as it is by any unexceptionable boundaries designated in nature. When the species of a vegetable order are exceedingly numerous, and a close similarity pervades the whole; genera are multiplied by botanists, that the discrimination of species may be facilitated. On the other hand, where a group of species is not unwieldy from its size, or deficient in distinctive marks, the genera are made as comprehensive, as natural affinity will permit. The diversity of structure, which exists in the flowers of *Gentiana*, or the fruit of *Bunias*, would be deemed ample foundation for constructing half a dozen genera among the umbelliferous, leguminous, or gramineous orders. But as the species of the genera above have a strong agreement in one part of their fructification, as well as in general habit, and as no great obscurity or inconvenience results from keeping them together, it has not been thought worth while to multiply nomenclature by arranging them under separate titles.

The separation of Gillenia from Spiræa is one of those cases, upon which the botanist may hesitate long, without finding reasons strong enough to influence his decision. The natural order to which they belong is remarkable for having its genera well defined, so that there is no necessity for the separation, arising from confusion or indistinctness. The fruit of Gillenia is exactly the fruit of Spiræa, and the habit of the herb in one is not very foreign from that of the other. There is nevertheless something in the irregular corolla, taken in conjunction with the campanulate calyx, which I think would prevent any one, at first sight, from considering the plant a Spiræa; and which may afford sufficient ground for following the example of Mœnch in considering it a distinct genus.

The Gillenia trifoliata grows in woods, in a light soil, from Canada to Florida. In the maritime states I have not met with it north of the Hudson. Its flowering time is in June and July.

The generic character, which distinguishes this plant from Spiræa, is as follows : *Calyx campanulate, five toothed ; corolla irregular, petals lanceolate, contracted near the claws ; capsules five*, The species *trifoliata* has *ternate, lanceolate, serrate leaves, and stipules which are minute, linea-lanceolate* and *nearly entire.*

Class *Icosandria*, order *Pentagynia.* Natural orders *Senticosæ*, Lin. *Rosaceæ*, Juss.

This plant has commonly a number of stems from the same root, which are a foot or two in height, erect, slender, flexuous, smooth, commonly of a reddish tinge, and considerably branched. The leaves are alternate, trifoliate, with very short petioles, furnished with small lanceolate, slightly toothed stipules at the base. Leafets ovate, lanceolate, acuminate, sharply toothed, the upper ones often single. The flowers are few in number, scattered, terminal, nodding, forming a sort of panicle, with long peduncles, occasionally furnished with minute lanceolate bractes. Calyx subcampanulate, or tubular, with the lower half narrowest, the border divided into five reflexed acute teeth. Petals five, the two upper ones separated from the three lower, white, with a reddish tinge on the edge of the outside, lanceolate, unguiculate, contracted, and approximated at base. Stamens about twenty in a double series within the calyx. Germ round, styles approximated. Capsules five, not *one*, as some authors have stated, diverging, oblong, acuminate, gibbous without, sharp edged within, two valved, one celled, one or two seeded; seeds oblong, corresponding in shape to the capsule.

The root of this plant is much branched and knotty. It consists of a woody portion, invested with a thick bark, which when dry is brittle, and very bitter to the taste. The predominant, soluble ingredients in this root appear to be a bitter extractive matter, and resin. When boiled in water, it imparts to it a beautiful, deep red, wine colour, and an intensely bitter taste. This decoction undergoes no change from alcohol or gelatine, though it gives a precipitate with muriate of tin. Water distilled from the root has its peculiar flavour, with little of the bitterness. A large portion of resin is precipitated on the addition of water to an alcoholic tincture of the root.

Under the name of Spiræa trifoliata, this plant is well known to students of the American Materia Medica, as an emetic. To the remarks which have been made by various writers, I can add my own testimony of its possessing properties in a certain degree analogous to those of ipecacuanha. It requires, however, a larger dose, and I have not been satisfied that it is at all certain in its operation. At times I have known fifteen grains to produce a full operation; at others thirty grains have been given to a person already predisposed to vomit, without producing the least sensible effect.

The best printed account which I have found respecting its mode of operation is contained in an Inaugural Dissertation, published at Philadelphia in 1810, by Dr. De la Motta, then of Charleston, S. C. This gentleman, in addition to other trials, took the root himself twice in sufficient quantity to produce vomiting. "In order," he says, "to ascertain this particular power of the Spiræa, I, early in the morning, fasting, prescribed for myself twenty-five grains of the powdered root of this plant. I divided this quantity into four equal parts, one of which I took every fifteen minutes, conceiving this a sufficient length of time to allow for the action of each dose in my stomach. The first dose taken produced no manifest effect. At the expiration of fifteen minutes I took a second dose;—a degree of uneasiness was experienced, attended with some nausea;—at the end of fifteen minutes more I swallowed a third dose,—nausea increased, until the convulsive action of my stomach took place. The fourth dose was now taken; considerable efforts were made to vomit, and finally the contents of my stomach were thrown up, together with a profuse quantity of bile. The determination of blood to my head, the frequency of my pulse, and heat of my system were much augmented. I now drank half a pint of warm water; the action of my stomach subsided, and the nausea gradually wore off. A portion of the medicine, I was induced to believe, had insinuated itself into the intestines, as two copious evacuations were produced within the space of three hours. During the day I felt much debilitated, but imputed this to the general effect of emetics.

"I was thus satisfied with respect to its efficacy as an emetic upon an empty stomach. But, being still desirous of becoming better acquainted with its particular operation after eating an usual meal, I made a second experiment, one month after the first. In the morning, one hour after I had eaten a hearty breakfast, I took twenty grains of the medicine, in divided doses, as in the former experiment. At the expiration of a very few minutes nausea commenced, which continuing to increase, with very few efforts I discharged the contents of my stomach. The effects of the second trial answered exactly my expectations."

Some authors have attributed a tonic power to the Gillenia, when administered in small doses. That it possesses such a power is rendered probble by its bitter taste, and by the fact, that small doses of ipecacuanha exert a beneficial stimulus on the stomach in certain cases of debility in that organ.

BOTANICAL REFERENCES.

Gillenia trifoliata, MŒNCH, *Meth. suppl. p.* 286.—NUTTALL, *Genera*, i. 307.—Spiræa trifoliata, LIN.—WILLD. *Sp. pl.* ii. 1063.—CURTIS, *Bot. Mag. t.* 489.—MILLER, *Icones*, 256.—MICHAUX, *Flor.* i. 294.—PURSH, i. 243.

MEDICAL REFERENCES.

SCHŒPF, 80.—B. S. BARTON, *Coll.* 26.—DE LA MOTTA, *Inaugural Dissertation.*

PLATE XLI.

Fig. 1. *Gillenia trifoliata.*
Fig. 2. *Calyx.*
Fig. 3. *A petal.*
Fig. 4. *Flower opened, shewing the situation of the stamens.*
Fig. 5. *Germ and styles.*
Fig. 6. *Styles separated.*

Rhus radicans

Annin

RHUS RADICANS.

Poison Ivy.

PLATE XLII.

Like the Rhus vernix, described in our first volume, this plant is regarded with aversion, and too frequently furnishes cause to be remembered by persons of susceptible constitution, who unwarily become exposed to its poisonous influence. The general recognition of its deleterious character is evinced in the application of the names *Poison vine*, *Poison creeper*, and *Poison Ivy*, which are given to it in all parts of the United States.

The Rhus radicans is a pretty common inhabitant of the borders of fields and of woods in most soils which are not very wet. Its mode of growth is much like that of the common creeper, the Ampelopsis quinquefolia of Michaux; and like that vine, and the European Ivy, it would doubtless be cultivated for ornament, were it harmless as it is handsome. As its name implies, this vine ascends upon tall objects in its neighbourhood by means of strong lateral rooting fibres, which project in great numbers from its sides, and attach themselves to the bark of trees and the surface of stones. The extreme branches of these fibres appear very strong in proportion to their fineness, and insinuate themselves into the minutest pores and crevices. The adhesion of the vine to the bark of trees is frequently so strong, that it cannot be torn off without breaking, and I have repeatedly seen large stems of the Rhus completely buried in the trunks of old trees, the bark having grown over and enveloped them. The fibres are analogous in their structure to fine roots, and consist of a regular wood and bark. They are sometimes thrown out in such numbers on all sides, as to give the vine a shaggy appearance and conceal its bark. In general, however, they tend to the shady side, and are attracted toward opaque objects, furnishing an exemplification of Mr. Knight's beautiful explanation of motion in tendrils, which, by their propensity to avoid the light and approach the shade, are directed into contact with bodies capable of yielding them support.

The size of the stem in this vine is commonly not more than an inch. Sometimes, however, in very old plants, it is found several times as large. It is usually compressed on the side which adheres to the support. In favourable situations it ascends to the tops of the highest rocks and trees; and is often seen restoring to decayed trunks the verdure which they have lost. When it does not meet with an elevated prop, the plant becomes stunted in its growth, is more branched, and affects a spiral mode of growth; or falls to the ground, takes root and rises again.

The genus Rhus is placed by Linnæus in the class *Pentandria*, and order *Trigynia*. The present species, however, is diœcious, a fact which is also true of most of the American species of Rhus which I have examined. The Rhoes belong to the Linnæan natural order *Dumosæ*, and to the *Terebintaceæ* of Jussieu.

The leaves of the Rhus radicans are ternate, and grow on long semicylindrical petioles. Leafets ovate or rhomboidal, acute, smooth and shining on both sides, the veins sometimes a little hairy beneath. The margin is sometimes entire and sometimes variously toothed and lobed, in the same plant. The flowers are small and greenish white. They grow in panicles or compound racemes on the sides of the new shoots, and are chiefly axillary. The barren flowers have a calyx of five erect, acute segments, and a corolla of five oblong recurved petals. Stamens erect with oblong anthers. In the centre is a rudiment of a style.—The fertile flowers, situated on a different plant, are about half the size of the preceding. The calyx and corolla are similar but more erect. They have five small, abortive stamens and a roundish germ surmounted with a short, erect style, ending in three stigmas. The berries are roundish and of a pale green colour, approaching to white.

A plant has long appeared in the Pharmacopœias under the name of *Rhus toxicodendron.* Botanists are not agreed whether this plant is a separate species from the one under consideration, or whether they are varieties of the same. Linnæus made them different with the distinction of the leaves being naked and entire in Rhus radicans, while they are pubescent and angular in Rhus toxicodendron. Michaux and Pursh, whose opportunities of observation have been more extensive, consider the two as mere local varieties; while Elliott and Nuttall still hold them to be distinct species. Among the plants which grow abundantly around Boston, I have frequently observed individual shoots from the same stock having the characters of both varieties. I have also observed that young plants of Rhus radicans frequently do not put out rooting fibres until they are several years old, and that they seem, in this respect, to be considerably influenced by the contiguity of supporting objects.

The wood of the Poison Ivy is brittle, fine grained and white, with a yellow heart in the old plants.

If a leaf or stem of this plant be broken off, a yellowish milky juice immediately exudes from the wounded extremity. After a short exposure to the air, it becomes of a deep black colour and does not again change. This juice, when applied to linen, forms one of the most perfect kinds of indelible ink. It does not fade from age, washing, or exposure to any of the common chemical agents. I have repeatedly, when in the country, marked my wristband with spots of this juice. The stain was at first faint and hardly perceptible, but in fifteen minutes it became black, and was never afterwards eradicated by washing, but continued to grow darker as long as the linen lasted.

Dr. Thomas Horsfield, in his valuable dissertation on the American species of Rhus, made various unsuccessful experiments with a view to ascertain the nature of this colouring principle, and the means of fixing it on stuffs. He found that the juice, expressed from the pounded leaves, did not produce the black colour, and that strong decoctions of the plant, impregnated with various chemical mordants, produced nothing more than a dull yellow, brownish or fawn colour. The reason of this is, that the colouring principle resides not in the sap, but in the *succus proprius* or peculiar juice of the plant, that this juice exists only in small quantity, and is wholly insoluble in water, a circumstance which contributes to the permanency of its colour, at the same time that it renders some other medium necessary for its solution.

With a view to ascertain the proper menstruum for this black substance, I subjected bits of cloth stained with it, to the action of various chemical agents. Water, at various temperatures assisted by soap and alkali, produced no change in its colour. Alcohol, both cold and boiling, was equally ineffectual. A portion of the cloth, digested several hours in cold ether with occasional agitation, was hardly altered in appearance. Sulphuric acid reddened the spots, but scarcely rendered them fainter. The fumes of oxymuriatic acid which bleached vegetable leaves and bits of calico in the same vessel, exerted no effect on this colour.

Boiling ether is the proper solvent of this juice. A piece of linen spotted with the Rhus was immersed in ether and placed over a lamp. As soon as the fluid boiled, the spot began to grow fainter, and in a few minutes was wholly discharged, the ether acquiring from it a dark colour. The linen at the same time became tinged throughout with a pale greyish colour, acquired from the solution.

This nigrescent juice, in common with that of the Rhus vernix, has, perhaps, claims to be considered a distinct proximate principle in vegetable chemistry.

The leaves and bark are astringent to the taste, which quality appears to be occasioned by gallic acid rather than tannin. The infusion and decoction become black on the addition of salts of iron, but discover hardly any sensibility to gelatin.

A poisonous quality exists in the juice and effluvium of this plant, like that which is found in the Rhus vernix already described. It is said, that some other species of Rhus, such as Rhus pumilum and Rhus typhinum, possess the same property in a greater or less degree. The poison Ivy, however, appears to be less frequently injurious than the poison Dogwood, and many persons can come in contact with the former with impunity, who are soon affected by the latter. I have never, in my own person, been affected by handling or chewing the Rhus radicans, though the Rhus vernix has often occasioned a slight eruption. Indeed, the former plant is so commonly diffused by road-sides and near habitations, that its ill consequences must be extremely frequent, were many individuals susceptible of its poison.

Those persons who are constitutionally liable to the influence of this poison, experience from it a train of symptoms very similar to those which result from exposure to the Rhus vernix. These consist in itching, redness and tumefaction of the affected parts, particularly of the face; succeeded by blisters, suppuration, aggravated swelling, heat, pain, and fever. When the disease is at its height, the skin becomes covered with a crust, and the swelling is so great as in many instances to close the eyes and almost obliterate the features of the face. The symptoms begin in a few hours after the exposure, and are commonly at the height on the fourth or fifth day; after which, desquamation begins to take place, and the distress, in most instances, to diminish.

Sometimes the eruption is less general, and confines itself to the part which has been exposed to contact with the poison. A gentleman, with whom I was in company, marked his wristband with the fresh juice, to observe the effect of the colour. The next day his arm was covered with an eruption from the wrist to the shoulder, but the disease did not extend further. It sometimes happens that the eruption continues for a longer time than that which has been stated, and that one set of vesications succeeds another, so as to protract the disease beyond the usual period of recovery.

The symptoms of this malady, though often highly distressing, are rarely fatal. I have nevertheless been told of cases in which death appeared to be the consequence of this poison.

The disease brought on by the different species of Rhus appears to be of an erysipelatous nature. It is to be treated by the means which resist inflammation, such as rest, low diet, and evacuations. Purging with the neutral salts is peculiarly useful, and in the case of plethoric constitutions, or where the fever and arterial excitement are very great, blood-letting has been found of service.

The extreme irritability and burning sensation may be greatly mitigated by opium. Cold applications, in the form of ice or cold water, are strongly recommended by Dr. Horsefield in his treatise, and when persevered in, they appear to exert a remarkably beneficial effect. The acetate of lead is perhaps as useful as any external palliative, and it should be used in solution rather than in the ointment, that it may be applied as cold as possible. The late Dr. Barton speaks highly of a solution of corrosive sublimate externally applied in this disease, but from trials of the two remedies made at the same time and in the same patient, I have found the lead the more beneficial of the two.

A person who has been in contact with the Rhus and finds himself poisoned, should immediately examine his hands, clothes, &c. to see if there are no spots of the juice adhering to him. These, if present, should be removed, as they will otherwise serve to keep up and extend the disorder. From a want of this precaution, the disease is frequently transferred from the hands to different parts of the body, and likewise kept up for a longer time than if the cause had been early removed. As washing does not eradicate the stains of this very adhesive juice, it is best to rub them off with some absorbent powder.

The Rhus radicans has been administered internally in certain diseases by a few practitioners in Europe and America. Dr. Horsefield, in several instances, administered a strong infusion in the dose of about a teacup full to consumptive and anasarcous patients. It appeared to act as an immediate stimulant to the stomach, producing some uneasiness in that organ, also promoting perspiration and diuresis. Some practitioners in the Middle States, we are told by the same author, have exhibited it with supposed benefit in pulmonary consumption. A French physician, Du Fresnoy, has reported seven cases of obstinate herpetic eruption, which were cured by the preparations of this plant. His attention was drawn to the subject by finding that a young man who had a *dartre* upon his wrist of six years' standing, was cured of it by accidentally becoming poisoned with this plant. The same physician administered the extract in several cases of palsy, four of which, he says, were cured by it.

Dr. Alderson, of Hull, in England, gave the Rhus toxicodendron in doses of half a grain, or a grain three times a day, in several cases of paralysis; and states, that all his patients recovered, to a certain degree, the use of their limbs.

The first symptom of amendment was an unpleasant feeling of prickling or twitching in the paralytic limbs. Dr. Duncan, author of the Edinburgh Dispensatory, states, that he had given it in larger doses without experiencing the same success, although he thinks it not inactive as a medicine.

My own opinion is, that the plant under consideration is too uncertain and hazardous to be employed in medicine, or kept in apothecaries' shops. It is true, that not more than one person in ten is probably susceptible of poison from it. Yet, even this chance, small as it is, should deter us from employing it. In persons not constitutionally susceptible of the eruptive disease, it is probably an inert medicine, since we find that Du Fresnoy's patients sometimes carried the dose as high as an ounce of the extract, three times a day, without perceiving any effect from it.

It is true that the external application of the Rhus radicans and Rhus vernix would, in certain cases, afford a more violent external stimulus, than any medicinal substance with which we are acquainted. But since it is neither certain in its effect, nor manageable in its extent, the prospect of benefit, even in diseases like palsy and mania, is not sufficient to justify the risk of great evil.

BOTANICAL REFERENCES.

Rhus radicans, WILLD. *Sp. pl.* i. 1481.—ELLIOTT, i.—Rhus toxicodendron, &c. MICHAUX, *Flor.* i. 183.—PURSH, i. 205.—Toxicodendron rectum &c.—DILLENIUS, *Elth. t.* 291.

MEDICAL REFERENCES.

DU FRESNOY, *quoted in Annals of Medicine*, iv. 182.—v. 483.—Med. and Phys. Journal, i. 308.—vi. 273.—x. 486.—DUNCAN, *Dispens.* 294.—HORSEFIELD, *Dissertation, Philad.* 1798.

PLATE XLII.

Fig. 1. *Rhus radicans, the barren plant in flower,*
Fig. 2. *Fruit.*
Fig. 3. *Barren flower.*
Fig. 4. *Fertile flower.*
Fig. 5. *Petal.*
Fig. 6. *Stamens and rudiment of a style in the barren flower,*
Fig. 7. *Germ, style and abortive stamens in the fertile flower.*

MYRICA CERIFERA.

Wax Myrtle.

PLATE XLIII.

Almost every region of the United States produces varieties of the Wax myrtle. Michaux considers them all as belonging to one species, a conclusion which is warranted by the great number of intermediate sizes, and forms of leaf, which may be observed between the different extremes. Pursh, however, has chosen to distinguish three species which bear wax, and which he names *cerifera* after Linnæus, *Caroliniensis* from Willdenow, and *Pennsylvanica* from Lamarck. The *Wax myrtle* or *Bayberry*, as it is often called, which is common in New England, varies in height from one to seven or eight feet. It is found in every kind of soil from the borders of swamps to the tops of barren hills, and is very much influenced in its size and appearance, by the place in which it happens to grow.

III

Myrica cerifera

The genus Myrica belongs to the class *Diœcia* and order *Tetrandria.* It is also ranked among the *Amentaceœ* of Linnæus and Jussieu.

The generic character consists in an *imbricated ament; the scales without a corolla; the barren flowers containing four anthers, the fertile ones two styles. Fruit, one seeded.*—The specific character, as given by Michaux, is as follows. *Leaves wedge-lanceolate, with a few serratures at top; barren aments lax; fruit spherical, naked, distinct.*

The Wax myrtle is found bearing fruit at every size, from the height of one foot, to six or eight. In Louisiana, it is said to grow to twelve feet. The top is much branched, and covered with a grayish bark. The leaves are wedge-lanceolate, varying in width, sometimes entire, but more frequently toothed, particularly toward the end. They are somewhat pubescent, a little paler beneath, and generally twisted, or revolute in their mode of growth. They are inserted in a scattered manner by short petioles. The flowers appear in May before the leaves are fully expanded. The barren ones grow in catkins, which are sessile, erect, about half an inch or three quarters long; originating from the sides of the last year's twigs. Every flower is formed by a concave rhomboidal scale, containing three or four pairs of roundish anthers on a branched footstalk. The fertile flowers, which grow on a different shrub, are less than half the size of the barren ones, and consist of narrower scales, with each an ovate germ, and two filiform styles.

To these aments succeed clusters or aggregations of small globular fruits resembling berries, which are at first green, but finally become nearly white. They consist of a hard stone inclosing a dicotyledonous kernel. This stone is studded on its outside with small black grains resembling fine gun-powder, over which is a crust of dry white wax, fitted to the grains and giving the surface of the fruit a granulated appearance. Botanically speaking, this fruit has been improperly called a berry, and a drupe; since it is always dry and never invested with a cuticle, or any thing but the grains and wax.

Every young part of the Wax myrtle has a fragrant, balsamic smell, which it communicates to the fingers when rubbed by them. This appears to be derived from a resinous exudation, which may be seen in minute points of a bright transparent yellow, covering the young shoots and under surface of the leaves. In the berries this resinous substance is within the wax.

The bark and leaves of the Myrica cerifera contain gallic acid, tannin, resin, and a small quantity of mucilage, which are manifested by their usual tests.

The wax of the Myrica is obtained for common purposes by boiling large quantities of the berries in kettles with water enough to cover them to the depth of several inches. The berries, which float at first, gradually subside to the bottom when the wax is melted off, which latter substance floats on the surface. When the boiling has been continued long enough to divest the berries of most of their wax, the liquid is suffered to cool, and the wax concretes on the top. It is purified by melting it anew, and is then cast into masses.

In this state it is of a greenish gray colour, with a consistence which is intermediate between that of bees wax and tallow, being brittle and not remarkable for adhesiveness or unctuosity. It burns with a white flame, which is less vivid than that of tallow or whale oil.

The chemical properties of this wax have been examined by M. Cadet, in France, and Dr. Bostock, in England. From their experiments, we learn that water has no action on the Myrtle wax, either cold or at the boiling heat. Dr.

Bostock informs us that alcohol, at the ordinary temperature of the atmosphere, has no action upon it; but one hundred parts by weight of the fluid, when boiling, dissolve about five parts of the wax. About four fifths of this is deposited by cooling, and the rest is slowly deposited in a few days, or may be precipitated by water. Of the mass of wax, a certain portion remains insoluble in alcohol.

Sulphuric ether, according to Dr. Bostock, dissolves but little of the wax, when cold, but acts upon it rapidly, when boiling, taking up somewhat more than one quarter of its own weight. Upon evaporation, the wax is deposited in a crystalline or lamellated form, its texture resembling that of spermaceti.—Rectified oil of turpentine, when assisted by heat, dissolves about six per cent of its own weight, most of which is deposited on cooling.—Pure potash, in water, renders the wax colourless by boiling, and forms a soap with a small part, which may be decomposed by an acid, and affords the wax nearly unchanged.—The sulphuric acid, assisted by heat, dissolves about one twelfth of its own weight, and forms a dark brown mass. The nitric and muriatic acids exert very little action upon it. Dr. Bostock considers the Myrtle wax to be a fixed vegetable oil, rendered concrete by oxygen.

M. Cadet, in addition to many of the above characteristics of Myrtle wax, found that it combined readily with the semivitreous oxyde of lead, forming a very hard plaister. When distilled in a retort, the wax was partly decomposed, and a portion which passed over was white and of a soft consistence. Oxygenated muriatic acid bleaches it, but with more difficulty than bees wax.

The experiments which I have made on this substance confirm the preceding statements with the following exceptions. Cold alcohol dissolves a minute portion, which is gradually separated by the addition of water, and floats in perceptible *flocci*, near the surface. Cold ether dissolves about one sixteenth of its weight. This it does with great rapidity, and if thin shavings of the wax be dropped into a vessel of ether, they disappear almost immediately.

Dr. J. F. Dana has published, in Silliman's Journal, an account of some experiments made to ascertain the proportion of wax, and of the other parts which compose the entire berry. He found the wax to constitute nearly a third of the whole, or thirty two per cent; the kernels 47.00, the black powder 15.00, with about 5.00 of a resino-extractive matter.

There undoubtedly exists, in the berries of this shrub, some interesting constituents beside the wax and insoluble portions, as the following results will show. If water be distilled from the fresh berries, it acquires a slight pearly appearance and a fine aromatic odour and taste. This indicates the presence of a volatile oil, though I have not performed the experiment sufficiently in the large way to obtain any oil separate. The decoction remaining in the retort gives proofs of gallic acid.

When the wax, in a separate state, is boiled in alcohol, a portion is dissolved, which is mostly deposited on cooling, leaving the fluid clear. But if alcohol be boiled upon the berries till a strong solution is formed, it does not give a deposit on cooling, but the solution coagulates into a soft solid and remains afterwards unaltered. This coagulum is readily soluble in cold ether, and melts when exposed to heat. If the berries be boiled in water until the wax is melted and principally detached, the remaining parts still give a coagulating solution with alcohol.—The tincture made by digesting cold alcohol on the bruised berries is considerably coloured, and becomes turbid on the addition of water, but whether the resinous substance thus precipitated is the same in small quantity, which produces the coagulation in a large one; I am not prepared to say.

It appears, then, that there exists in the berries of the Myrica a peculiar vegetable principle, bearing nearly the same relation to alcohol, as starch and gelatin do to water. I have not yet obtained it in a separate state, and cannot therefore give any additional characteristics to those which have been already stated.

The Myrtle wax is useful for many of the purposes for which bees wax and tallow are employed, particularly for candles. It burns with a clear flame, though less vivid than that of common oil, and emits a considerable fragrance. It was formerly much in demand as an ingredient in a species of blacking ball, to which it communicated a temporary lustre and power of repelling water. It has occasionally been used in pharmacy in various compositions intended for external use, and is mild or stimulating according as it is more or less pure and freed from the colouring matter.

In some parts of Europe plantations of this shrub have been raised with a view to the profit to be derived from the wax. In this country, where the shrub abounds, the berries are often neglected, their collection and the separation of the wax being deemed too laborious to compensate the trouble.

In Dr. Thatcher's Dispensatory, we are informed, on the authority of Dr. Mann, that the bark of the root of the Myrica cerifera is emetic. With a view to examining thoroughly its medicinal properties, Dr. S. L. Dana, in 1818, made it the subject of an inaugural dissertation. He found that the powdered bark was acrid and astringent, but did not appear to possess any other qualities than were attributable to those two. Moderate doses of the powder and decoction produced no effect on the stomach or bowels. Large doses, for instance two scruples, were swallowed with difficulty on account of their acrimony, and occasioned heat and nausea at the stomach. Larger doses, of a drachm, produced a powerful burning sensation and vomiting. Costiveness generally followed the use of this medicine. The powder, when snuffed up the nose, proved powerfully sternutatory, and when applied to the fungous granulations of an ulcerated leg, it produced so much pain as compelled the patient to wash it off.

We may then consider the bark of the Myrica as an acrid stimulant and astringent. That it sometimes proves emetic, in large doses, is to be explained in the same way as the operation of mustard and horse-radish, which some of the older physicians employed as emetics. When the stomach is burdened with an undue quantity of stimulus, it naturally tends to relieve itself by vomiting.

On the whole, we are to esteem the Myrica cerifera as more interesting in a chemical, than a medical point of view. The pleasant aroma of the water distilled from the berries, and the application of the wax to some purposes of pharmacy, are all, that this shrub at present offers, much deserving the attention of physicians.

BOTANICAL REFERENCES.

Myrica cerifera, Willd. iv. 745.—Michaux, ii. 227.—Pursh, ii. 620.—Myrtus foliis lanceolatis, &c.—Gronovius, 155.—Myrtus brabanticæ similis, &c.—Catesby, i. 13?

MEDICAL AND CHEMICAL REFERENCES.

Cadet, *translated in Nicholson's Journal, 8vo. vol.* iv. 187.—Bostock, in ditto, 129.—Kalm, *Travels,* i. 129.—Dana, *in Silliman's Journal,* vol. i.—Thacher, *Disp.* 288.

PLATE XLIII.

Fig. 1. *Myrica cerifera, with fruit not fully ripe.*

Fig. 2. *A barren branch in flower.*

Fig. 3. *Fertile ditto in flower.*

Fig. 4. *A barren flower.*

Fig. 5. *The same with the scale turned down, shewing the mode of growth of the anthers.*

Fig. 6. *Fertile flower.*

Fig. 7. *Fruit somewhat magnified.*

PL. XLIV.

Juniperus communis

JUNIPERUS COMMUNIS.

Common Juniper.

*PLATE XLIV.**

THE prostrate variety of the common Juniper is so peculiar in its mode of growth, that it has some claims to be considered a distinct species. On comparing it, however, with European specimens, I find the similarity so great, that I do not see sufficient grounds for separating it, especially as there are, in Europe, several varieties in size and mode of growth, which are not recognized as separate species. The variety, which is the only one I have met with in the Northern States, is a large trailing shrub, continually throwing out roots from its branches, and spreading in all directions until it forms beds, which are many rods in circumference. In this way it continues to advance outward, supporting itself by new roots even after the original trunks, at the centre,

* Very beautiful drawings, from which this and the following plate are engraved, were sent me by a lady in Hampshire county.

are dead and decayed. It seldom rises more than two or three feet from the ground.

The genus *Juniperus* belongs to the class *Diœcia*, order *Monadelphia*, and natural order *Coniferæ* of Linnæus and Jussieu. It is distinguished by an *ovate ament with peltate scales, which, in the barren flowers, are whorled in threes, with from two to four anthers; in the fertile ones opposite. Berry three seeded.*—In the common Juniper, *the leaves are ternate, spreading, mucronate, larger than the berry.*

The Juniper is with us always a shrub, never rising into a tree. The tips of the branches are smooth and angular. The leaves grow in threes and are linear-acerose, sharply mucronate, shining green on their lower surface, but with a broad glaucous line through the centre of the upper. These leaves, however, are always resupinate, and turn their upper surface toward the ground. The barren flowers grow in small axillary aments, with roundish, acute, stipitate scales, inclosing several anthers. The fertile flowers, growing on a separate shrub, have a small, three parted calyx growing to the germ; and three styles. The fruit is a fleshy, roundish, oblong berry, of a dark purplish colour, formed of the germ and confluent calyx, marked with three prominences or vesicles at top, and containing three seeds. It requires two seasons to arrive at maturity from the flower.

The leaves of the Juniper have a strong and rather unpleasant taste, with a little astringency. The peculiar juice of the bark appears to consist of resin and volatile oil. Gum Sandarach, which furnishes the material of pounce, is obtained from the European Juniper, from which it exudes spontaneously through crevices and perforations in the bark.

The part principally used in medicine is the berries. These have a strong peculiar taste, accompanied with considerable sweetness. When long chewed, they leave an impression of bitterness. The sweetness appears to reside in the pulp, the bitterness in the seeds, or in their immediate investment, and the aromatic flavour in the essential oil. Dr. Lewis observes, that tinctures, made with these berries, differ according as they are prepared with the berries entire or bruised.

When of a good quality these berries yield, in distillation, a large quantity of pungent, volatile oil of a peculiar flavour, the same which it communicates to gin. The medicinal powers, for which this article is employed, may be considered as residing in this oil.

The berries of the Juniper have long been employed for the purposes of a diuretic, particularly in dropsy. Many of the older writers, whose names are of high authority in medicine, have given favorable reports of the operation of this medicine in hydropic cases. It has been used in substance, in infusion, and in various compound medicines. The effects of its most popular preparation, that of an ardent spirit, are too universally known to require particular elucidation. In addition to the specific effect of the essential oil, some physicians have attributed virtues to the rob, or inspissated decoction of the berries. Hoffman found it of great use in debility of the stomach and intestines, particularly in old people. The stronger preparations have been found useful in uterine obstructions, and in paralytic affections of the vesica urinaria.

Linnæus informs us, in his Flora Lapponica, that a fermented decoction of Juniper berries is used in Sweden as a common drink, but he denies the infusion being substituted for tea and coffee, as some writers have stated, in Lapland. Woodville and others have misquoted him on this point.

The American Juniper berries are considerably inferior to the European in strength and flavour. The best of the latter are said to be from Italy. But of the imported specimens, which I have examined at the druggists' shops in this country, very few possess any remains of the original strength, and much the greater portion of them appear to have undergone at least one distillation, before their exportation from Europe. The best Juniper berries have a strong impregnation of volatile oil, which, having been once tasted, cannot be easily mistaken. Those which have been subjected to distillation are dry and tasteless.

BOTANICAL BEFERENCES.

Juniperus communis, Linn.—Smith, *Flor. Brit.* iii. 1085.—*Engl. Bot. t.* 1100.—Woodville, ii. *t.* 95.—Michaux, ii. 245.—Pursh, ii. 646.—Blackwell, *t.* 187.

MEDICAL REFERENCES.

Murray, *App. Med.* i. 34.—Lewis, *Disp.* 240.—Linnæus, *Flora Lapp.* 376.—Woodville, *ut supra.*

PLATE XLIV.

Fig. 1. *A branch of Juniperus communis in fruit.*

Fig. 2. *A barren twig in flower.*

Fig. 3. *Barren ament.*

Fig. 4. *Scale of anthers of the same.*

Fig. 5. *Fertile flower.*

Pl. XIV.

Juniperus virginiana

JUNIPERUS VIRGINIANA.

Red Cedar.

PLATE XLV.

Unlike the subject of the preceding article, this species rises into a tree of considerable size. It is the largest of the Junipers growing within the original limits of the United States, though it appears that Lewis and Clarke brought home specimens of a lofty tree, with foliage resembling the Savin, found on the banks of streams among the Rocky mountains, and which is supposed to be the same with J. excelsa, growing in Siberia.

Michaux, in his North American Sylva, informs us, that it is found from Maine and from Lake Champlain, without interruption to the Cape of Florida. In the Middle and Northern states, it frequents the most barren soils, being found in abundance upon dry, rocky hills, where scarcely any other tree can subsist. Its size, however, is said to be here inferior to that which it attains in Virginia, and farther south.

Its habit and foliage abundantly distinguish it from the species in the last article. From the Juniperus Sabina, or common Savin of Europe, its botanical distinction is by no means easy. The general appearance and arrangement of the leaves in the full grown specimens of both is precisely the same, except that in the Red cedar the leaves are shorter and more compactly imbricated, having an ovate form, while in the Savin they are somewhat longer and more remote, and may be called lanceolate. In the Red cedar they are also more universally and pungently acute. The characters of the latter species, which I have seen given by different botanists, are almost all defective, in ascribing to it ternate leaves, which, I believe, never exist, except in imperfect or distorted specimens. Its more true character is as follows. *Trunk arboreous, upper leaves imbricated in four rows, ovate, pungently acute.* It is by no means certain that on mature examination all the present species of Juniper will be found sufficiently distinct to be kept separate.

The Red cedar, when full grown, is a middling sized tree, though, on account of the value of its wood, it is seldom suffered to reach its full dimensions. The trunk is straight and decreases rapidly from the ground, giving off many horizontal branches. Its surface is generally unequal, and disfigured by knots, and the crevices and protuberances they occasion. The small twigs are covered with minute, densely imbricated leaves, which continue to increase in size as the branch grows, till they are broken up and confounded with the rough bark. These leaves are fleshy, ovate, concave, rigidly acute, marked with a small depressed gland on the middle of their outer side, growing in pairs, which are united at base to each other, and to the pairs above and below them. They do not alter their situation, but continue opposite till they are obliterated by age. A singular variety sometimes appears in the young shoots, especially those which issue from the base of the trees. This consists in an elongation of the leaves to five or six times their usual length, while they become spreading, acerose, considerably remote from each other, and irregular in their insertion, being either opposite or ternate. These shoots are so dissimilar to the parent tree that they have repeatedly been mistaken for individuals of a different species.—The barren flowers grow in small oblong aments, formed by peltate scales with the anthers concealed within them. The fertile flowers have a proper perianth, which coalesces with the germ and forms a small round-ish berry, with two or three seeds, covered on its outer surface with a bright blue powder.

The leaves of the Red cedar have a strong disagreeable taste, with some pungency and bitterness. The peculiar taste and odour reside, no doubt, in a volatile oil, which, however, is not readily separated by distillations in a small way. The tincture becomes turbid when water is added, and very much so if suffered to stand a day or two. The presence of tannin is developed by the admixture of dissolved isinglass, with a decoction of the bark and leaves.

The botanical similarity of this tree to the *Savin*, which is an European shrub, has already been mentioned. In their sensible and medicinal properties, they are equally allied. The taste of the two species is nearly the same, except that the cedar leaves are the more nauseous of the two. As the American tree is frequently known throughout the country by the name of Savin, our apothecaries have been led to presume upon its identity with that medicine, and it has long been used in cases where the true Savin is recommended. Its most frequent use, however, is in the composition of the cerate employed for keeping up the irritation and discharge of blisters.

This preparation is the same with the Savin cerate, used in Europe, the leaves of the Red cedar being substituted for those of the Savin. When properly prepared by boiling the fresh leaves for a short time in about twice their weight of lard with the addition of a little wax, a cerate is formed of peculiar efficacy as a perpetual epispastic. When applied as a dressing to a newly vesicated surface, and afterwards repeated twice a day, it rarely fails to keep up the discharge for an indefinite length of time. Under its operation, the discharge usually changes from a serous to a puriform appearance, and concretes upon the surface; so that it requires to be removed from time to time, to admit the full action of the cerate.

Internally the leaves of the Juniperus Virginiana have been found to exert effects very similar to those of the Savin. They have proved useful as an emmenagogue, and as a general stimulant and diaphoretic in rheumatism. They have also had some reputation as a diuretic in dropsy.

The wood of the Red cedar is smooth, light, and very durable. Its alburnum is white, but the heart wood of a beautiful red colour, whence its name is derived. It is principally employed for posts in fences, in which capacity it proves more durable than almost any species of wood used for the same purpose.

BOTANICAL REFERENCES.

Juniperus Virginiana, WILLD. iv. 852.—PURSH, ii. 647.—MICHAUX, *fil. N. A. Sylva, t.* 155.

MEDICAL REFERENCE.

THACHER, *Disp.* 247.

PLATE XLV.

Fig. 1. *Juniperus Virginiana in fruit.*
Fig. 2. *Variety with long leaves.*
Fig. 3. *Barren branch in flower.*
Fig. 4. *Barren ament magnified.*
Fig. 5. *Scale and anthers.*

Pl. XII

Menyanthes trifoliata

MENYANTHES TRIFOLIATA.

Buck Bean.

PLATE XLVI.

The *Buck bean* or *Marsh Trefoil* is one of those plants which are native in Europe and North America, with so little difference of structure, in the two continents, that their specific identity can hardly be doubted. I have compared specimens of the native, and foreign plant, without being able to perceive the least definable difference, except in size; the American being smaller. Yet, if we admit the statements of botanical writers, the plant flowers in England at least a month later than it does in the neighbourhood of Boston, a circumstance not usual in other species of vegetables.

The most spongy and boggy soils, which are inundated at certain seasons, and never wholly destitute of water, are the favorite situations of the Menyanthes trifoliata. It often constitutes large beds at the margin of ponds and brooks. It is common in New England, and grows, according to Pursh, as far south as Virginia.

The genus *Menyanthes* has its *corolla hairy on the upper side; stigma bifid; capsule one celled, two valved.* The species in the present article is named from its *ternate leaves.* Class *Pentandria.* Order *Monogynia.* Natural orders *Rotaceæ*, Lin. *Gentianæ*, Juss.

The root of this plant penetrates horizontally in the bog-earth to a great distance. It is regularly intersected with joints at the distance of about half an inch from each other, these joints being formed by the breaking off of the old petioles and their sheaths. The leaves proceed from the end of the root on long stalks furnished with broad sheathing stipules at base. They are trifoliate, nearly oval, glabrous, somewhat fleshy, and slightly repand, or furnished with many irregularities at the edge, which hardly prevent them from being entire. The scape is round, ascending and smooth, bearing a conical raceme of flowers. Peduncles straight, scattered, supported by ovate concave bractes. Calyx erect, subcampanulate, five parted, persistent. Corolla funnel shaped, the tube short, the border five cleft, spreading and at length revolute, clothed on the upper part with a coating of dense, fleshy, obtuse fibres. The colour, in the American variety, is generally white, with a tinge of red, particularly on the outside. Stamens five, shorter than the corolla, and alternate with its segments; the anthers oblong-arrow shaped. Germ ovate; style cylindrical, persistent, as long as the corolla; stigma bifid, compressed. Capsule ovate, two valved, one celled. Seeds numerous, minute, attached to two lateral receptacles.

In New England this plant flowers about the middle of May.

The whole plant and particularly the root has an intensely bitter taste, hardly exceeded by that of Gentian and Columbo. This bitterness resides chiefly in an extractive matter, soluble in water and spirit. The root is, however, resinous and impregnates alcohol more strongly than water, and may be precipitated from its tincture, in part, by the latter fluid.

The root of this vegetable is undoubtedly entitled to a high place in the list of tonics. In Europe it has long been admitted to a place in the Materia Medica, and has received the commendations of various physicians. When given in small doses, about ten grains, it imparts vigour to the stomach and strengthens digestion. Its tincture, moderately used, has the same effect. Large doses, such as a drachm of the powdered root, or two or three gills of the saturated decoction, produce vomiting and purging, and frequently powerful diaphoresis. In this respect it agrees with many vegetable bitters, and perhaps resembles most nearly the Eupatorium perfoliatum. Its bulk, however, and unpleasant taste render it inconvenient to be used as an evacuant.

We are told by authors that the Buck bean has been employed with benefit in intermittent and remittent fevers. Boerhaave, in his own case of gout, was relieved by drinking the juice of the plant mixed with whey. Other physicians have found it useful in keeping off the paroxysms of that complaint. Dr. Cullen informs us, that he has "had several instances of its good effects in some cutaneous diseases of the herpetic or seemingly cancerous kind. It was taken by infusion in the manner of tea." Others have commended this vegetable in rheumatism, dropsy, scurvy and worms. Its reputation in the North of Europe, particularly in Germany, was at one time so high, that it was consumed in large quantities, and deemed a sort of panacea. Its true character, however, is simply that of a powerful bitter tonic, like Gentian and Centaury, to which it is closely related in its botanical habit, as well as sensible properties.

We may regard this plant as one of the numerous vegetable bitters abounding in our country, which are fully equal in strength to imported articles of their class, and which may hereafter lessen our dependance on foreign drugs.

Linnæus, in his Flora Lapponica, informs us, that in times of scarcity, sheep will subsist upon this plant, notwithstanding its bitterness. The Laplanders employ it as a substitute for hops to prevent acescency in their beer. They even introduce it in some instances into their bread, upon which Linnæus bestows the epithet, "amarus et detestabilis."

BOTANICAL REFERENCES.

Menyanthes trifoliata, LINN. *Sp. pl.*—ŒDER. *Flora Dan. t.* 541.—CURTIS, *Flor. Lond.* 4. *t.* 17.—WOODVILLE, *Med. Bot. t.* 2.—SMITH, *Engl. Bot. t.* 495.—MICHAUX, *Flora.* i. 125.—PURSH, i. 139.—Menyanthes palustre triphyllum, RAY. *Syn.* 285.—Trifolium paludosum, GERARD. *em.* 1194.

MEDICAL REFERENCES.

MURRAY, *Apparatus med.* ii. 33.—LINNÆUS, *Fl. Lap.* 50.—HALLER, *Hist. Stirp. Helv.* 633.—CULLEN, *Mat. Med.* ii. 53.—THOMPSON, *Lond. Disp.* 256.

PLATE XLVI.

Fig. 1. *Menyanthes trifoliata.*
Fig. 2. *Calyx.*
Fig. 3. *Petal.*
Fig. 4. *Stamen.*
Fig. 5. *Style.*
Fig. 6. *Fruit.*

Ranunculus bulbosus

RANUNCULUS BULBOSUS.

Bulbous Crowfoot.

PLATE XLVII.

It is a remarkable fact that a great portion of the *weeds*, which are most troublesome in the United States, are of European origin, having introduced themselves since the discovery of this country. Some of these emigrants have settled in our grazing and mowing lands, such as the Ranunculus bulbosus, acris and repens, indiscriminately called *Buttercups*, *Crowfoot*, and *Yellow weed;* the Chrysanthemum leucanthemum, or *White weed;* the Rumex acetosella, or *Sorrel;* the Hypericum perforatum, or *St. John's wort*, &c. In our cornfields and gardens are quartered the *Couch grass*, Triticum repens ; the different species of *Goosefoot* or *Pig weed*, Chenopodium ; the *Dock*, Rumex crispus, &c. ; the *Charlock* or *Wild Radish*, Raphanus Raphanistrum ; *Burdock*, Arctium lappa, &c. Some have commenced their inroads within a few years, such as the Cnicus arvensis, improperly called *Canada thistle*; the Genista tinctoria or *Dyer's weed*, &c.—In return for these introductions, we have sent them the Erigeron Canadense, and the prolific families of Ambrosia and Amaranthus.

No race of plants is more familiarly known than the Ranunculi. Of numerous species, both native and imported, which we possess; several resemble each other so nearly, as to pass with common observers for the same plant. The great similarity of their properties renders it almost unnecessary in a medical or economical point of view to distinguish them. I have selected the bulbous-rooted species, not because it is more active in its properties than many others, but because it is one of the most common and best known.

The genus belongs to the class *Polyandria*, and order *Polygynia*. It is found in the natural orders *Multisiliquæ*, Linn. and *Ranunculaceæ*, Juss. Its generic character is formed by *a five leaved calyx; five petals, with a melliferous pore at the base of each; the seeds naked.* No genus can be more strictly natural than this. A general resemblance pervades the whole of the species, which indicates their consanguinity at sight.

The nectary, the never failing concomitant of this genus, is a small cavity at the inside of the claw of each petal, generally covered by a flat scale, sometimes surrounded with a concave brim, and at others inclosed in a short cylinder. A subtle and violent acrimony, on which the medical properties seem to depend, is found in most, if not in all, of the species.

The species *bulbosus* has *compound leaves, an erect many flowered stem, a furrowed peduncle, reflexed calyx, and bulbous root.* It grows generally in dry pastures, mowing lands and road sides, flowering abundantly in May and the first part of June, after which it gives place to its equally abundant successors, R. acris and repens, which, however, generally prefer a more moist soil. These three species, having flowers of similar size and appearance, are indiscriminately known by the name of *Buttercups.* Their distinction affords a pleasing instance of different combinations of features, forming separate characters for similar plants. The *R. bulbosus* has a furrowed flower-stalk and reflexed calyx; *R. repens* a furrowed flower-stalk and spreading calyx, and *R. acris* a round flower-stalk and spreading calyx.

A more particular description of the plant in our figure is as follows. Root fleshy, solid, roundish, depressed, sending out radicles from its under side. In autumn it gives off lateral bulbs near its top, which afford plants for the next year, while the old root decays. Stems several, erect, round, hairy, branching. Root leaves on long petioles, ternate, sometimes quinate; the segments variously cut, lobed and toothed; hairy. Stem leaves sessile, ternate, the upper ones more simple. Flowers several on a stem, solitary, of a bright glossy yellow. Peduncles furrowed, angular, hairy. Calyx leaves oblong, hairy, bent back against the peduncle. Petals five, inversely heart shaped; the nectary at the claw covered with a small wedge-shaped emarginate scale. Stamens numerous, yellow, with oblong erect anthers. Germs numerous with reflexed stigmas. Fruit a spherical head composed of acute, naked, diverging seeds with recurved points.

The roots of Ranunculus bulbosus appear to consist principally of albumen intermixed with ligneous fibres. If the root be macerated in cold water, it gives a solution of this substance, which coagulates in flocks on the application of heat; and undergoes the same process slowly on the admixture of alcohol. But the most interesting constituent in this, and in most other species, is the acrid principle which pervades every part of the plant in its green state. Like the acrimony of the Arum, it is volatile, and disappears in drying, or upon the application of heat. It differs, however, in not being destroyed by a moderate heat, and in being fully preserved in distillation. I have subjected various species of Ranunculus to this experiment, and always found the distilled water to possess a strong acrimony; while the decoction and portions of the plant remaining in the retort were wholly destitute of this property. This distilled water, when first taken into the mouth, excited no particular effect; but after a few seconds a sharp, stinging sensation was always produced. When swallowed, a great sense of heat took place in the stomach. I preserved some of the water distilled from leaves of Ranunculus repens, for several months in a close stopped phial; during which time it retained its acrimony undiminished. In winter time it froze, and on thawing had lost this property. Tilebein, as quoted by Dr. Pulteney, in some experiments on this genus of plants, found that water distilled from *R. sceleratus*, on cooling, deposited small crystals, which were hardly soluble in any menstruum, and were of an inflammable nature. I have not met with an appearance of this kind. The distilled water, however, had a substance dissolved in sufficient quantity to yield a gradual precipitate with some reagents, such as muriate of tin and acetate of lead. The strength of the distilled water is impaired by continuing the operation too long. The acrimony of the plant is expended in a very short time at the boiling heat, and a farther continuance of the distillation brings over only water.

Since the time of Dioscorides [*Note* A.] the acrid and stimulating properties of the Ranunculi have been well known. This acrimony resides in all the species, with the exception of *R. auricomus*, which is said to be mild, and perhaps two or three others. It is so powerful that it speedily inflames or corrodes the lips and tongue, if kept in contact with them. In the nostrils it acts as a violent sternutatory, and if swallowed in considerable quantity, it brings on great pain, heat and inflammation of the stomach, and has even occasioned convulsions and death.

Before the introduction of Cantharides as a vesicatory, different species of Ranunculus were used upon the skin, as external stimulants. Their power of occasioning erosion and ulceration appears to have been known to the ancients. Different medical writers have given accounts of their mode of operation; but the most extensive history and investigation is that of Krapf, published at Vienna, in 1766. This work, which I have not seen, is quoted in all its principal facts by Professor Murray of Göttingen in the Apparatus medicaminum. According to this author the various species, with which his experiments were made, proved capable of exciting inflammation, blistering and ulceration, when applied to the skin. A slice of the fresh root of R. bulbosus placed in contact with the inside of the finger, brought on a sense of burning in two minutes. When taken off, the skin was found without redness, and the sense of heat and itching ceased. In two hours, however, it returned again, and in ten hours a full serous blister was raised. This was followed by an ulcer of bad character and difficult to heal. He remarks that, if the application is continued after the first itching, the pain and subsequent erosion is much greater.

From the accounts given of this species, also of R. sceleratus, R. acris, and some others, it appears that the leaves, flowers, buds, or roots of these plants, if bruised and applied to the skin, excite redness and vesication. This effect is not constant, but fails to take place in certain constitutions or at certain seasons of the year. Generally, however, they are said to operate in half an hour, or less, from the time of their application. They are stated to possess the advantage over blisters made by flies, that they never occasion symptoms of strangury.

With a view to their external stimulus they have been used advantageously in rheumatism, the hip disease, hemicrania, and fixed pains of various descriptions. Among the old practitioners, who have recorded instances of their effects, are Baglivi, Storck, and Sennertius. A curious practice, at one time, prevailed in several countries in Europe, of applying the Ranunculus to the wrists or fingers, for the cure of intermittent fever. This is mentioned by Van Swieten, Tissot, and some others. In hemicrania it was applied to the head, and in this case it did not produce a discharge, nor break the skin; but occasioned tumefaction of the hairy scalp.

An objection against the use of the Ranunculi, as external stimulants, exists in the uncertainty of their operation, and the violent effects which sometimes have followed after they had been applied. Those writers, who have witnessed their application, record instances in which these vegetable blisters have been followed by deep, ill-conditioned and sloughing ulcers, which were not healed without great difficulty. Tissot mentions an instance, in which an application made to the thumb caused a deep, painful ulcer, which penetrated to the bone, and occupied some months in its cure. In another case the blister spread, in a few hours, over the whole arm, occasioning fever and delirium, and was followed by such a tendency to gangrene, that the limb was with difficulty saved. Chesnau, quoted by Murray, advises that the Ranunculus should be applied to a small surface only, and through a perforation in an adhesive plaister, to prevent it from spreading. From want of this caution, he had known extensive inflammation to arise and spread over a greater part of the face, neck, and breast.—Linnæus, in his Flora Suecica, relates that beggars, in Sweden, were known to excite ulcerations of their feet with the Ranunculus sceleratus, to assist them in extorting charity from passengers.

I know not to what extent the efficacy of the Ranunculi, externally applied, can be depended on. Certain it is that they do not affect all persons alike, and this fact is avowed by those who have used them most. I have repeatedly made applications of the contused roots and leaves of different species to my arm and hand; and worn them for a dozen hours, without feeling any particular sensation, or perceiving any visible effect. The rapid drying up of the moisture of the plant seemed to prevent it from acting upon the skin. I am inclined to believe, there is something in the action of these vegetables analogous to that of the poisonous species of Rhus described in this work; which some individuals, but not all, are susceptible of. The extensive and spreading inflammation, which they occasionally produce, resembles more the effect of these shrubs, than of any of the ordinary rubefacients or vesicants.

The burning sensation which the Ranunculi excite in the mouth when chewed, extends to the stomach if they are swallowed. Krapf states that a small portion of a leaf or flower of R. sceleratus, or two drops of the juice, excited acute pain in the stomach, and a sense of inflammation in the throat. He gave a large quantity of the juice to a dog, which brought on vomiting and great distress; and the animal being killed, was found with the stomach inflamed and contracted, and the pylorus hardly pervious. The same author informs us that dilution greatly diminishes the power of this fluid, so that half a drachm of the juice, in six ounces of water, may be taken with entire safety.

Dr. Withering, as quoted by Dr. Pulteney in the Linnæan transactions, asserts, that the distilled water of Ranunculus flammula is an emetic more instantaneous and less offensive than sulphate of zinc. I know not in what publication of Dr. W. this statement is made, but the fact appears to me not improbable. Acrid substances, such as mustard, pepper, and horse-radish, if swallowed in large quantities, excite the stomach to relieve itself by vomiting. An objection, however, exists against the distilled water, owing to the uncertainty of its strength; which must vary in proportion to the quantity of the plant employed, the time occupied in distillation, and the subsequent time for which the fluid is kept.

Krapf states that R. auricomus and R. lanuginosus are so free from acrimony, that they are eaten as greens or sallads. All the species lose their pungency in boiling, so that even the R. sceleratus, one of the most acrid, is used for the same purpose.

Grazing cattle generally avoid the plants of this genus, which grow among grass, as far as it is possible for them to do it. Accordingly we observe the flowers of Ranunculi left untouched, while the grass is closely cropped around them. It is nevertheless unavoidable, so common are these plants, that portions of them should be eaten very often by these animals. It is probable that small quantities of the less acrid sorts do them no injury. At least, it appears that their stomachs are much less susceptible to this kind of stimulus than ours. In the *Pan Suecus* some experiments upon these plants, with domestic animals, are detailed; in which, it is stated that, horned cattle refused to eat all the species when offered to them, except R. auricomus. This species was rejected by horses, while they would eat R. flammula. Sheep and goats eat the R. acris, one of the most pungent species. Dr. Pulteney states, as a well known fact, that hogs, in England, devour the roots of R. bulbosus. How it is that these animals resist the deleterious effects of so virulent plants, it is not easy to say. It is, however, a not more remarkable fact, than the power of some animals to devour Cantharides and even mineral poisons with impunity.*

In their dry state, various species of Ranunculus enter into the composition of hay, particularly R. acris. Having lost their acrimony altogether in drying, they are harmless and probably nutritive.

* See a note, vol. i. p. 164.

Dr. Pulteney has published a memoir in the Linnæan transactions on the economical use of some of the Ranunculi, particularly the R. fluviatilis, which he considers a variety of R. aquatilis. Contrary to the common effects of the other species, this plant is said, by him, to be not only innocent, but highly nutritive to cattle. He states that, "in the neighbourhood of Ringwood, on the borders of the Avon, which affords this vegetable in great abundance all the year, some of the cottagers sustain their cows, and even horses, almost wholly upon this plant; since the remaining part of their food is nothing more than a scanty pittance, they get on the adjacent heath, which affords little more than Ling, Lichen, Bog-moss or Sphagnum, &c. It is usual to employ a man to collect a quantity for the day every morning, and bring it in the boat to the edge of the water, from which the cows, in the instance seen, stood eating it with great avidity. I was indeed informed," says he, "they relished it so highly, that it was unsafe to allow them more than a certain quantity; I think between twenty five and thirty pounds daily, each; but with variation according to circumstances. The cows I saw were apparently not in a mean condition, and gave a sufficient quantity of good milk. I was told by the person whose cattle were feeding on it, that he kept five cows and one horse so entirely on this plant and what the heath afforded, that they had not consumed half a ton of hay throughout the whole year; none being used except when the river was frozen over. I examined the whole parcel on which four cows were feeding, in the beginning March, and found the whole consisted exclusively of the Ranunculus fluviatilis without any mixture of the Potamogeton, Carex, Sparganium, or other aquatic plants. In summer, however, it can hardly be avoided but that there must be a mixture of some of these, but other plants are not chosen.

"This account was confirmed to me by different persons; by whom I was further informed that hogs are also fed with the same plant, on which they improve so well, that it is not necessary to allow them other sustenance, till it is proper to put them up to fatten."

In Veterinary practice the Ranunculus bulbosus has been employed as an external stimulant. To this purpose Dr. Chapman, in his Therapeutics, thinks it may be better adapted than other topical excitants.

BOTANICAL REFERENCES.

Ranunculus bulbosus, Linn. *Sp. pl.*—Curtis, *Flora Lond.* i. 38. —Martyn, *Flora rustica, t.* 28.—Smith, *Flora Britt.* 591.—*Engl. Bot. t.* 515.—Michaux, i. 321.—Pursh, ii. 393.—Ranunculus tuberosus magor, J. Bauhin, iii. 417.—Ranunculus pratensis, &c.—Œder, *Fl. Dan. t.* 515.

MEDICAL REFERENCES.

Murray, *Apparatus, Med.* iii. 88.—Krapf. *Ranunculi. Vienna,* 1766.—Lewis, *Mat. Med.* ii. 262.—B. S. Barton, 23.—Pulteney, *Lin. transactions,* v. 14.—Chapman, *Therapeutics,* ii. 411.

PLATE XLVII.

Fig. 1. *Ranunculus bulbosus, the radical leaf of the largest size and more subdivided than common.*

Fig. 2. *Radical leaf of smaller size and more common shape.*

Fig. 3. *Petal and nectary.*

Fig. 4. *Two stamens enlarged.*

Fig. 5. *Fruit.*

ILLICIUM FLORIDANUM.

Starry Anise.

PLATE XLVIII.

The same qualities which entitle the Liriodendron and Magnolias to a place among medicinal plants, exist abundantly in the kindred genus of Illicium. This family consists of fine, spicy, flowering shrubs, one of which, the I. anisatum, growing in Eastern Asia, derives its name from the similiarity of its flavour to that of Anise, a quality which exists, though less simple, in the subject of the present article. Another species, the I. parviflorum, a shrub with small yellowish flowers, first discovered by Michaux in the mountains of Georgia and Carolina, has so exactly the flavour of the Sassafras root, that they are not to be distinguished by the taste. The I. Floridanum forms beautiful thickets in the country bordering on the north of the Gulf of

Pl. XLVIII.

Illicium Floridanum

Mexico, and is often mentioned by the traveller Bartram, with his accustomed enthusiasm, as one of the chief beauties of that exuberant region. In the Northern states, as well as in Europe, it is preserved by artificial heat. The drawing, which illustrates our description, was made from a greenhouse specimen.

The character assigned to this genus is formed by *a six leaved calyx, twenty seven petals, and a number of capsules arranged in a circle, two valved, one seeded.* The species *Floridanum* has its *leaves acuminate and its petals numerous, oblong and linear.*

The class and order are *Polyandria, Polygynia;* and the Natural orders *Coadunatæ*, Linn. *Magnoliæ*, Juss.

The Illicium Floridanum is a shrub, in some instances entitled to be considered a small tree. Its leaves are scattered, or grow in tufts, on short petioles. They are evergreen, oval lanceolate, slightly acuminate, entire, smooth on both sides, and firm or fleshy. The flower buds proceed from the sides of the branches at the axils of the last year's leaves. The flowers grow on slender, nodding peduncles, an inch or two long. When fully expanded, they are about the size of a dollar, and of a dark, purplish crimson. Calyx deciduous. Petals linear, obtuse, in three rows, about nine in a row, the uppermost row ascending, the lowermost descending, and broader or more spatulate. Stamens thirty or more, diverging, flat, depressed with the anthers recurved; pollen white. Germs a dozen or more, roundish-rhomboidal, compressed and arranged in a circular manner; styles short, recurved, pubescent on the inside. The fruit, which I have not seen, is represented by authors, as has been stated in the generic character.

The leaves and young shoots of this species of Starry anise abound in a fine, clear mucilage, which becomes immediately perceptible in the mouth, if these parts are chewed, and which communicates to water in a short time a ropy consistence. This mucilage is separated from the decoction by alcohol in the form of dark brown, tough, stringy coagula. Muriate of tin causes a precipitate after these coagula are withdrawn, which seems to indicate the presence of extract. Sulphate of iron added to the decoction, coagulated the mucus and darkened the colour. I discovered no traces of resin in the portions submitted to experiment, and a strong tincture was not disturbed by water. The trial, however, was conducted on a small scale.

The bark and leaves of the Illicium Floridanum are strongly impregnated with a spicy, aromatic taste and smell, approaching that of the Magnolias and Liriodendron, but perhaps more similar to that of some of the pungent seeds, particularly Anise and Coriander, between which they seem intermediate. This aroma is preserved in the distilled water, and fills the room with its fragrance, while distillation is going on. I was not able in my limited experiments to separate any volatile oil or camphor, on one of which principles, as in similar cases, the aroma doubtless depends.

An account of this species of Illicium is given, with a figure, in the Philosophical transactions for 1770, by John Ellis, Esq. He says, "We are indebted for the discovery of this curious American tree to a servant of William Clifton, Esq. of West Florida, who was sent to collect specimens of all the rarer plants by his master; and in April 1765, he met with it growing in a swamp near Pensacola. After this, in the latter end of January 1766, Mr. John Bartram, the king's botanist for the Floridas, discovered it on the banks of the river St. John, in East Florida, as appears from his description of it, and a drawing of a seed-vessel with some of the leaves, sent to Mr. Collinson." Mr. Bartram's description of it, as it appears in his journal up the river St. John, published by Dr. Stork, in his account of East Florida, is as follows. "Near here my son found a lovely, sweet tree, with leaves like the sweet bay, which smelled like Sassafras, and produces a strange kind of seed-pop; but all the seed was shed. The severe frost had not hurt it;—some of them grow nearly twenty feet high, a charming bright evergreen aromatic."*

Of the medicinal properties of this shrub, I am unable to speak with the certainty, which might have attended an extensive number of trials, made with the bark of full grown specimens. From the evidence afforded by the bark and leaves of a greenhouse specimen, and by the analogy of other species, and similar trees, I should not feel much hesitation in attributing to the Illicium the properties of a tonic-stimulant and diaphoretic. I have at least satisfied myself that the bark of a twig, and three or four of the leaves, produce no unpleasant consequence. Its bitter taste and aromatic quality point out its analogy to Cascarilla, Canella, Sassafras, and other aromatic barks, which are regularly consumed in the shops. Its

* It is very possible the above description may have been intended for Illicium parviflorum.

co-species, the Illicium anisatum of the East, is used as a condiment to communicate an agreeable flavour to certain dishes. The Chinese chew it after dinner as a stomachic and a sweetener of the breath. In some parts of the East Indies the natives and Dutch mix it with their tea and sherbet. It is also burnt as incense before their idols by some of the oriental nations, and carefully kept as an antidote to various poisons.

The beauty of both these shrubs renders them desirable acquisitions to collectors of plants.

BOTANICAL BEFERENCES.

Illicium Floridanum, LINN.—CURTIS, *Bot. Mag. t.* 439.—MICHAUX, i. 526—PURSH, ii. 380.

MEDICAL REFERENCES.

ELLIS, *in Philosophical transactions abridged,* xiii. 87. *t.* 2.—SCHŒFF, 91.

PLATE XLVIII.

Fig. 1. *Illicium Floridanum.*
Fig. 2. *Several stamens magnified.*
Fig. 3. *Pistils magnified.*
Fig. 4. *A pistil separate.*

ARISTOLOCHIA SERPENTARIA.

Virginia Snakeroot.

*PLATE XLIX.**

It is probable that this root, like many other articles now used in medicine, was indebted to its sensible qualities, for its first introduction into use. As the name implies, its earliest medicinal character was founded on a supposed antidotal power against the bite of venomous serpents. Cornutus, at the end of his book on the plants of Canada, published at Paris in 1635, tells us, that a root had been sent to him from "*Notha Anglia,*" which was called *Serpentaria*, and in the vernacular tongue *Snagröel.* This root was a very sure safeguard against the bite of a huge serpent in that country, which proved inevitably fatal within twelve hours, unless a good portion of the antidote

* I am indebted to a gentleman in Georgia for the very natural drawing of this plant.

1
2
3
4
5

was swallowed in season; which being done, no one was ever known to be in danger of his life from this cause.

The snagröel has had a great many rivals in the character of specifics against the bite of serpents. So great, indeed, is the number of articles which are called uniformly successful in such cases, that we are compelled to believe, that the bite of the rattle snake, and doubtless of other venomous serpents in the country, although attended with severe and alarming consequences, is nevertheless but seldom fatal; and hence that the honor of proving specific in these cases is one of cheap acquisition.

The Serpentaria grows in woods in the Southern and Middle parts of the United States. It bears cultivation in any part of the union, though the most northerly situation, from which I have received wild specimens, is the vicinity of New Haven, from which place some living plants were sent to me by Dr. Monson.

The genus *Aristolochia* has *a monopetalous, tubular, crooked corolla, swelling at base, and dilated at the border. Capsule inferior, six celled.* The species *Serpentaria* has its *leaves heart-shaped, oblong, acuminate; stem flexuous; peduncles radical.* Pursh mentions a variety with leaves so narrow, as to appear like a distinct species; the flower, however, being not different. Woodville's figure of our plant has the leaves much too broad for the common habit of the vegetable.

It belongs to the class *Gynandria*, order *Hexandria*, or more properly *Dodecandria*. It is one of the few genera placed by Linnæus in that class which are not of the Orchideous tribe. Natural orders *Sarmentaceæ*, Linn. *Aristolochiæ*, Juss.

This vegetable is humble in its growth, being most commonly under a foot in height. The root is extremely fibrous, and sends up a number of stems. These are simple or slightly branched, jointed, flexuous, and often of a reddish tinge. The leaves are alternate, on short petioles, oblong, entire, acuminate, heart-shaped at base and three nerved.

The flowers grow close to the ground, like those of Asarum. They have a stiff leathery texture, and a dull brownish purple colour. The peduncle which supports them has one or more leafets, and gradually enlarges into a furrowed obovate germ. The corolla, like others in this singular genus, consists of a long contorted tube, bent in the form of the letter S, swelling at its two extremities, having its throat surrounded by an elevated edge or brim, and its border expanded into a broad irregular margin, forming an upper and under lip, which are closed in a triangular manner in the bud. The anthers are twelve in number, growing in pairs to the sides of the fleshy style, which is situated in the bottom of the corolla, and covered by a firm, spreading convoluted stigma, which extends over the anthers. The capsule is obovate, six angled, six celled, with numerous small flat seeds.

Snakeroot has a penetrating, rather agreeable, resinous smell, and a pungent bitter taste, resembling somewhat that of the Pinus Canadensis, or Hemlock spruce. It communicates its qualities both to spirit and water, but most to the former. I subjected a quantity of the root to distillation for one hour, and obtained in the receiver a whitish pearly fluid, very strongly impregnated with the aroma, but less bitter than the root. On standing twenty four hours, this fluid deposited round the edges of the surface a considerable number of small white crystals, which proved to be pure camphor. They were inflammable, fusible with a sudden, and volatile with a gradual heat. I perceived no essential oil, though Dr. Lewis informs us, that if the quantity of root, submitted to the operation, be large, there arises a small portion of pale coloured essential oil of a considerable smell but of no very strong taste. There is probably a portion of resin present, as I found that the root, after having been boiled in water an hour, still impregnated alcohol so as to cause a precipitate with water. The bitterness communicated to the infusion and decoction appears to reside in a variety of extractive matter.

Medicinally considered, Serpentaria is a tonic, diaphoretic, and in certain cases an antispasmodic and anodyne. It has been abundantly used in fevers of various descriptions, and has been commended by a host of medical writers. There is no doubt that it has been injudiciously employed in many cases, in fever attended with an active pulse and inflammatory diathesis. The early stages, also, of febrile diseases rarely admit the exhibition of so decided a stimulant, without injury. But in the advanced stages of fever and those attended with typhoidal symptoms, this medicine is resorted to with great advantage, both alone and in combinaton with other tonics and stimulants. It is peculiarly useful in supporting the strength, and in allaying the irregular actions which attend great febrile debility, such as subsultus tendinum, delirium, watchfulness, &c. Its bitter ingredients, and the camphor which it contains, no doubt contribute to these effects. It is most advantageously given in combination with bark, or with wine and opium.

Snakeroot is a popular remedy in exanthematous disorders as a diaphoretic, being given to keep out the eruption, and to restore it when it has receded. In the latter case its use is doubtless injudicious, and if it fails to reproduce the eruption, it greatly increases the heat, pain, and restlessness of the patient. It is better in cases where the eruption has receded to the disadvantage of the patient, to attempt its restoration by nauseating and saline diaphoretics, and even by a full emetic, than to incur the risk of aggravating the symptoms by a stimulating regimen.

Dr. Chapman, in his Therapeutics, considers the Serpentaria as partaking the mixed qualities of a stimulant and tonic, and acting also as a diaphoretic and diuretic. It is peculiarly useful as an auxiliary to the bark. He states, that one of the more early uses of the medicine was in the cure of intermittent fever. Whether alone it was found adequate to this purpose, does not clearly appear. "It was used by Sydenham in conjunction with wine, to prevent the recurrence of the paroxysm, and from his account, not without advantage. As a general rule, he says, that in all cases, where it is expedient to combine wine with bark, the effect will be much increased by adding Serpentaria. The correctness of this observation has been fully confirmed by subsequent experience, and it is now very much the practice to unite the two articles in the low states of disease."

Dr. Chapman farther states, that though it is doubtful whether the Serpentaria, by itself, will cure ague and fever, it is certainly a powerful assistant to bark, not only in increasing its efficacy, but, what is of great consequence, in enabling the stomach to retain the medicine.

To remittent fever he thinks this medicine better adapted. It has here, in many cases, an indisputable superiority over bark, inasmuch as it is rarely offensive to the stomach, and may be given without injury, in those obscure states of the disease, where the remission is not readily discernible. He prefers, in these cases, a combination of bark, snakeroot, and soda.

Snakeroot, he informs us, is much resorted to as a popular remedy in the management of the secondary stages of pleurisy. After bleeding, it is the ordinary practice, in many parts of our country, to resort to a strong infusion of this article with a view of exciting perspiration. Catarrhs, rheumatisms, and other winter complaints, incident to rustic life, are managed in the same way. In that species of pleurisy which is properly enough designated by the epithet bilious, he has repeatedly had occasion to recur to the Serpentaria, and always with more or less utility. This bilious pleurisy he considers as having all the characters of pneumonic inflammation, with the addition of some of the symptoms incident to autumnal fever, such as headach, great gastric distress, and almost always violent vomitings of bile. It differs also from ordinary pleurisy in having less activity of inflammation, and consequently in not bearing the same extent of depletion. The system, indeed, will often be very evidently depressed by one or two bleedings. In this case the practice which has been commonly pursued is, after the removal of a comparatively small portion of blood, and the thorough evacuation of the alimentary canal; to administer very freely draughts of the infusion of the Serpentaria in order to excite copious diaphoresis.

Dr. Chapman concludes his remarks on this article, by stating, that it is admirably suited to check vomitings, and to tranquillize the stomach,

12

more particularly in bilious cases. It is given for this purpose in decoction, in the small dose of half an ounce or less at a time, and frequently repeated.

The most common form of exhibiting snakeroot is in infusion, for which purpose half an ounce may be steeped in a pint of boiling water for two hours, in a covered vessel. Of this infusion an ounce or two may be taken every three or four hours. Decoction is a less proper mode of preparing this plant, as it tends to dissipate the volatile parts, a portion of which is detained in a state of mixture by the infusion. Sometimes the powder is given in doses of from ten to thirty grains. A tincture of snakeroot is made by digesting an ounce of the root in a pound or somewhat less of proof spirit. The compound tincture of bark, commonly called Huxham's tincture, contains Serpentaria as one of its ingredients.

BOTANICAL REFERENCES.

Aristolochia serpentaria, LINN. *Sp. pl.*—WALTER, *Flor. Car.* 223.—WOODVILLE, ii. 291. *t.* 106.—MICHAUX, ii. 162.—PURSH, ii. 596.—Pistolochia sive Serpentaria Virginiana, &c.—PLUKENET, *t.* 148. *f.* 5.—CATESBY, *Car.* i. 29.

MEDICAL REFERENCES.

MURRAY, *App. Med.* i. 348.—CULLEN, *Mat. Med.* ii. 85.—CHAPMAN, *Therapeutics*, ii. 411.—LIND. *Hot climates*, 104, 254.

PLATE XLIX.

Fig. 1. *Aristolochia serpentaria with the flower beginning to expand.*
Fig. 2. *Side view of the flower expanded.*
Fig. 3. *Front of ditto.*
Fig. 4. *Longitudinal section of the flower.*
Fig. 5. *Style, anthers, and stigma magnified.*
Fig. 6. *Fruit.*

ALETRIS FARINOSA.

Star Grass.

PLATE L.

I KNOW of no plant which surpasses the Alteris farinosa in genuine, intense and permanent bitterness. Neither aloes, gentian, nor quassia exceed it in the impression produced on the tongue. It has, on account of this property, attracted the observation of some medical men, and may hereafter become an article of more consequence in the Materia Medica. Although the number of trials, hitherto made, are perhaps not sufficient to fix with precision its exact character, yet in a collection of American medicinal vegetables it ought not to pass unnoticed.

This plant grows in most parts of the United States in fields and about the edges of woods, and flowers in June and July. I have found it near Boston on the south, but never to the north of it.

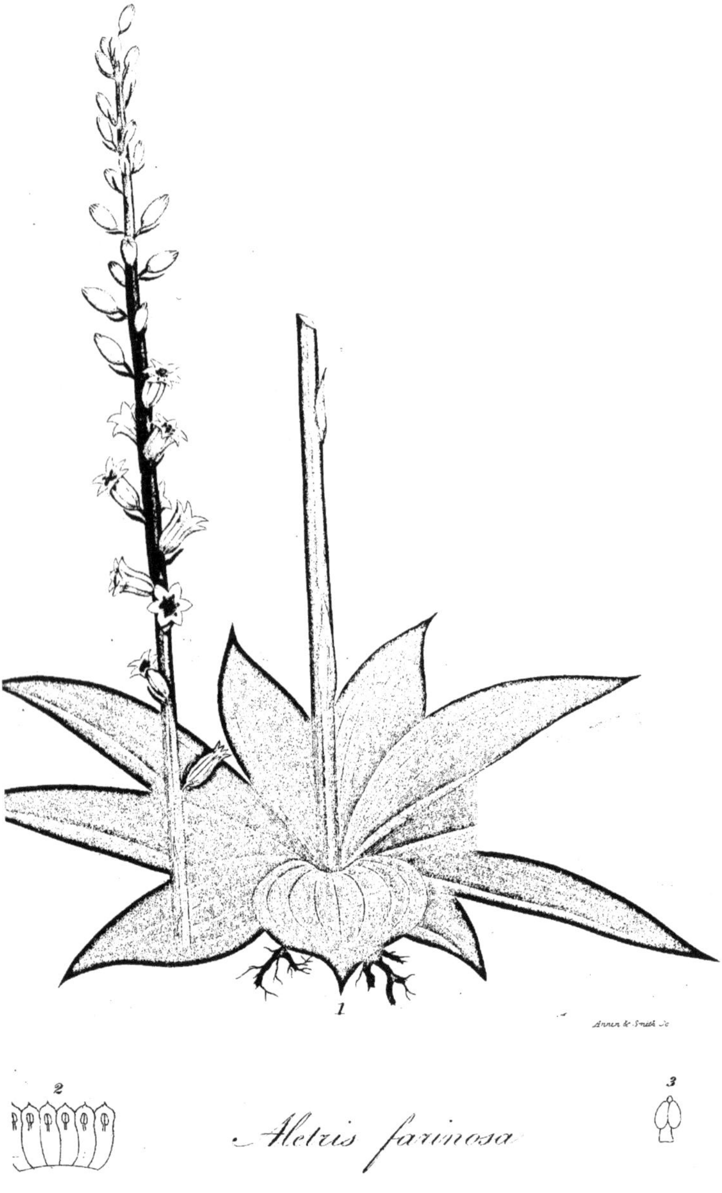

1

Annan & Smith Sc

2

Aletris farinosa

3

Its mode of growth is not without beauty, the leaves spreading close to the ground in a radiated manner, like a star ; while the spike is supported by an almost naked stalk, at a distance above them. The names *Star Grass* and *Blazing star* are generally given to it in the country, from the peculiar appearance of its leaves.

The genus Aletris has its *corolla tubular, six cleft, wrinkled, persistent ; stamens inserted into the base of the segments ; style triangular, separable into three ; capsule opening at top, three celled, many seeded.* The species *farinosa,* called *alba* by Michaux and Pursh, has its *flowers pedicelled, oblong-tubular, somewhat wrinkled in fruit ; the leaves broad lanceolate.* Michaux observes that of the species referred by Linnæus to this genus, the A. farinosa is the only one which strictly belongs to it. Class *Hexandria ;* order *Monogynia ;* natural orders *Liliaceæ,* Linn. *Asphodeli,* Juss.

This plant has a single circle of radical leaves, which are sessile, nerved, lanceolate, and smooth. The stem or scape is from one to three feet high, invested with remote scales, which sometimes expand into small leaves. The flowers form a slender, scattered spike with very short pedicels and minute bractes. Calyx none. Corolla white, of an oblong bell-shape, divided at the mouth into six acute, spreading segments. The outside, particularly as the flower grows old, has a roughish, wrinkled or mealy appearance, by which the specific name was suggested. The stamens are short, inserted near the mouth of the corolla at the base of the segments. The circumstance of their being opposite to the segments, and not alternate with them, affords the most distinguishing mark of this genus. The anthers are somewhat heart-shaped. Germ pyramidal, half inferior, tapering: style triangular, separable into three. Capsule invested with the permanent corolla, triangular, three celled, three valved at top. Seeds numerous, minute, fixed to a central receptacle.

The *Aletris aurea*, of Michaux and Pursh, closely resembles this species, and it is difficult, by comparing specimens of the two, to point out any permanent distinctive marks. The leaves of A. aurea are somewhat narrower and the flowers bright yellow. Walter places it under A. farinosa as a variety, and adds that he could not detect a specific difference; although the time of flowering and place of growth indicate that they are distinct. In sensible properties they are similar.

In the London Philosophical transactions for 1730, a plant is mentioned by Clayton, which, though not described in botanical language, leaves little doubt that the Aletris farinosa is intended. He says, "there is another root of the species of hyacinths; the leaves are grass-like, but smooth and stiff, of a willow-green colour, and spread like a star on the ground. From the middle shoots a tall, long, rush-like stem, without leaves, near two feet high; on one side grow little white bell-flowers one above another. The root is black outwardly, but brown within. It is bitter and probably has the same virtues as Little Centaury. Some call it *ague grass*, others *ague root*, others *star grass*."

The root of the Aletris is highly resinous, and appears to contain a portion of extractive matter. The tincture, made by digesting the root in alcohol, is intensely bitter, and assumes a milky turbidness if water be added to it. The decoction is moderately bitter, and is not disturbed by alcohol. With chalybeate solutions it undergoes little change. The tincture is to be considered a stronger preparation than the decoction, although the latter has a good share of the virtues of the plant.

The bitterness of this vegetable has brought it into notice in the quality of a tonic and stomachic. I have been informed of its use for this purpose by physicians in different parts of the country. The most common mode of its employment, I understand, is by infusion or decoction. Pursh speaks of it as a remedy in the colic, but on what principle it can operate in relieving that disease, I am at a loss to say.—The amount of bitter resin, which the plant contains, led me to suspect that it might possess some of the properties of aloes, to which the plant is botanically related; but on trial, made in several instances with the root in powder, a dose of ten or twelve grains produced no effect of this kind whatever. A physician, who experimented with larger quantities, with a view to test this quality, informed me that a dose of twenty grains occasioned much nausea and tendency to vomit, followed by some dizziness; but that no cathartic operation took place.

Dr. Cutler, in his account of the plants of New England, informs us, that this plant has been considered useful in chronic rheumatism; but does not mention the dose or preparation.

As far as we can sum up the testimony hitherto offered respecting the general properties of this plant, it appears that the infusion or decoction acts as a tonic in small doses. Indeed the exhibition of large ones would be inconvenient from the extreme bitterness of the plant. The powder, in small quantities, produces no immediate visible effect, except that it has appeared to invigorate the appetite. In large doses it disturbs the stomach, and possibly exerts some narcotic effect on the system. It remains to be determined whether these consequences are attributable to the resin, which the infusion does not dissolve; or whether the largeness of the dose is alone instrumental. It is well known that the stomach does not tolerate even gentian or any common bitter in large a dose. And it seems probable that if the Aletris should ever increase in reputation as a tonic bitter, it will only be by its use in limited quantities.

BOTANICAL REFERENCES.

Aletris farinosa, LINN.—WILLD. *Sp. pl.* ii. 183.—*Bot. Mag. t.* 1418.—Aletris alba, MICHAUX, *Flora.* i. 189.—PURSH, i. 225.—Hyacinthus floridanus spicatus, PLUKENET, *amalth.* 119, *t.* 437, *f.* 2.—Hyacynthus caule nudo, &c,—GRONOV. *Virg.* 38.

MEDICAL REFERENCES.

CLAYTON, *Phil. Trans. abr.* viii. 333.—CUTLER, *American Acad.* vol. i. 435.

PLATE L.

Fig. 1. *Aletris farinosa.*

Fig. 2. *Corolla opened to shew the insertion of the stamens.*

Fig. 3. *Stamen magnified.*

Fig. 4. *Pistil magnified.*

Pl. 11

AMERICAN

MEDICAL BOTANY.

RHODODENDRON MAXIMUM.

American Rose bay.

PLATE LI.

THE scenery of the American forest is distinguished not less by the greatness of its natural features, the imposing and picturesque appearance of its mountains, its rocky precipices, its broad streams and lakes; than it is by the magnificent clothing of wild shrubs and trees, the uncommon beauty of which, gives to rough and inaccessible spots a richness, that cultivation can hardly imitate. The Kalmia, described in our first volume, and the Rhododendron of the present article, which are reared with care and difficulty as ornaments of European gardens and pleasure grounds, can be seen in perfection no where but in the uncultivated recesses of our own continent.

Near the summits of mountains, on the banks of torrents and deep ravines, from which rivers take their rise, where the deep shade, moist soil and dashing water, preserve the atmosphere in a state of perpetual humidity; these shrubs, in luxuriant size and vigour, are seen to cover tracts of great extent, at one season presenting an unbroken landscape of gorgeous flowers, and at another with their evergreen foliage forming an impenetrable shelter for the wild animals of the forest.

Of the Rhododendron maximum, Mr. Pursh has designated three varieties. These are, 1. The Red, which inhabits swamps and the borders of mountain lakes from Canada to Carolina; 2. The White, found in the swamps of New Jersey and Delaware; 3. The Purple, on the highest mountains of Virginia and Carolina. This last variety is represented as peculiarly magnificent, growing to the size of a small tree, having its trunk eighteen inches and more in diameter, and its foliage triple the size of any other species.

The first variety of this elegant shrub grows abundantly on the banks of Charles river, a dozen or fifteen miles from Boston. It even supports the winter as far north as the state of Maine, and was observed, by Dr. Eaton, growing plentifully on the borders of Sebago lake near Portland. It does not bear transplantation well, but is apt to dwindle after the first or second year. It succeeds best when removed to a damp springy soil, and to a situation calculated to afford it shelter from the sun.

The Rhododendron, of the Northern states, is a large straggling shrub, very irregular in its mode of growth. The bark is of a greyish colour, very much cracked and broken. The leaves are in tufts at the ends of the branches. They are evergreen, coriaceous, on round fleshy petioles, oblong-oval, entire, revolute at the edges, and pale underneath. Both leaves and petioles, when young, are covered with a light woolly substance. The flowers form a terminal cluster or thyrsus immediately above the leaves, the stalks and calyces of which are covered with a glutinous pubescence. Previous to its expansion, the whole bunch forms a large compound bud, resembling a strobilus or cone, each individual flower-bud being covered by a rhomboidal bracte, which falls off when the flower expands. Calyx small, of five unequal obtuse segments. Corolla monopetalous, funnel-shaped, with a short tube, the border divided into five large, unequal segments, which are white, shaded with lake, the upper and largest, having a collection of orange coloured spots at its centre. Stamens declinate, unequal; the filaments white, thickened and hairy at base; anthers two celled, opening by two pores at top; pollen white. Germ ovate, hairy, glutinous; style declinate, equal to the longest stamens, thickened upwards; stigma a rough surface with five points. Capsule ovate, obtusely angular, five-celled. Seeds numerous, minute.

Considered in its chemical character, this shrub is a resinous astringent. A decoction of the leaves gives strong proofs of the presence of tannin in large quantities. Both the bark and leaves, digested in alcohol, produce a resinous tincture, which is immediately rendered turbid by water. The glutinous covering of the flower stalks appears of a resinous nature. A decoction of the leaves in water affords nothing which is not soluble in alcohol, and did not alter by it in two days' standing.

I have been induced to examine the Rhododendron and to insert it in this work, on account of the reputation it has possessed of being poisonous. The late Professor Barton, in his collections towards an American Materia Medica, has given various intimations of this sort, the most conclusive of which is his expression, "This is certainly a poison."—The result of my own attention to this shrub does not give reason for attaching to it suspicions of possessing a very deleterious nature. None of its external characters would lead to apprehensions of this sort, particularly the taste, which is simply astringent and herbaceous, and much like that of a common oak leaf. I know not what quantity might prove injurious, but under the conviction that the plant was not particularly dangerous, I have swallowed a green leaf of the middle size, so large that it required some resolution to masticate so unpalatable a morsel, but have found no ill effect whatever to result from it.

Medicinally considered, I think it must be ranked among the astringents, a place which both its sensible and chemical properties entitle it to hold. If it have any narcotic powers, they will probably be developed only by an extraordinary dose, which few persons will be likely to put to the test.

BOTANICAL REFERENCES.

Rhododendron maximum, Willd. *Sp. pl.* ii. 606.—*Bot. Mag. t.* 951.—Schmidt, *Arb. t.* 121.—Pursh, i. 297.—Michaux, *N. A. Sylva, t.* 67.

MEDICAL REFERENCE.

B. S. Barton, *Collections*, i. 18.

PLATE LI.

Fig. 1. *A branch of Rhododendron maximum in flower and in bud.*

Fig. 2. *Calyx and style.*

Fig. 3. *Stamen.*

PL. LII.

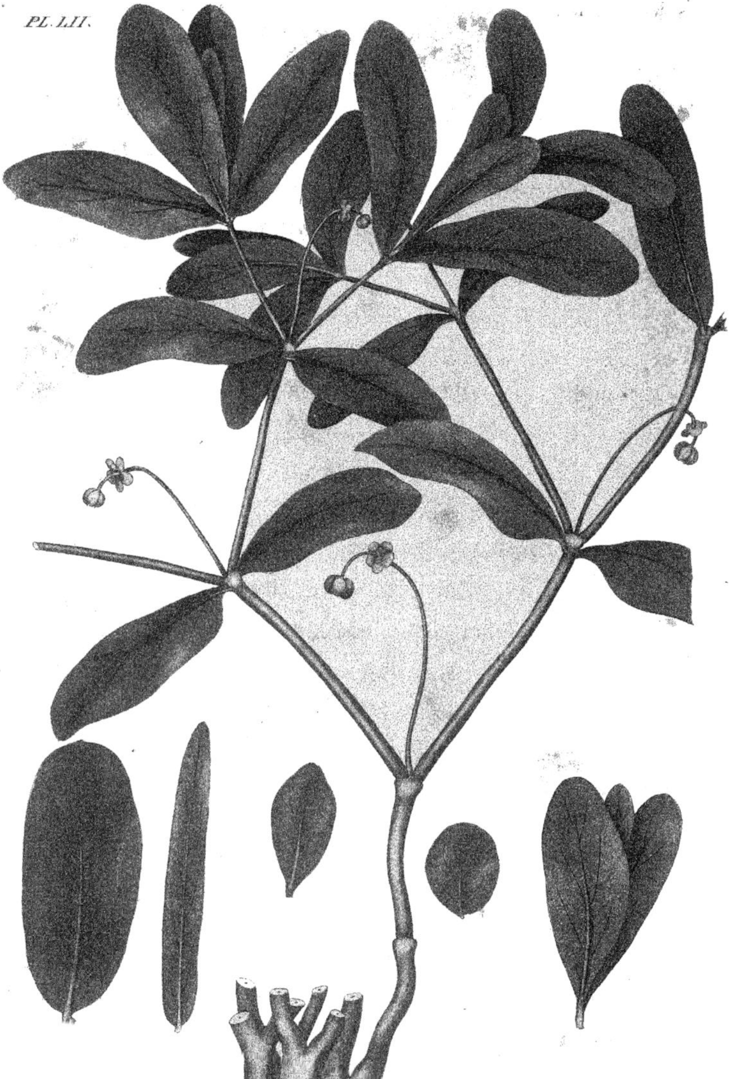

Euphorbia Ipecacuanha

EUPHORBIA IPECACUANHA.

Ipecacuanha spurge.

PLATE LII.

FROM the specific name given to this vegetable we infer, that before the true origin of the officinal ipecacuanha was known, this plant, among others, was for a time considered the source of that drug. The Pharmacopœia Danica was one of the works in which this reference was made, and Linnæus undoubtedly paid some respect to the opinion in assigning the specific name.

Nearly all the species of Euphorbia appear to possess the power of acting with violence on the stomach and alimentary canal. This power particularly resides in a milky juice which they exude on being wounded. Of the species which have been most extensively submitted to experiment are Euphorbia officinarum, esula, helioscopia, dulcis, peplus, exigua, Cyparissias, palustris, and Characias. Professor Murray has collected details respecting the operation of most of these, from various medical authorities. It appears that they all excite vomiting or purging, and in large doses bring on violent burning pains of the stomach and bowels, heat and thirst, followed by great prostration of strength, cold sweats, and in some instance, death. In small quantities, however, they have been used as medicines with safety, although some of them are uncertain in regard to their dose, and difficult to manage in their operation. [*Note* B.]

The genus Euphorbia comprises a vast number of species, of different habit, size and mode of growth. The flowers are frequently minute, very complex, and difficult of examination. They have a *calyciform involucrum with four or five segments like petals, and the same number of interior segments like nectaries. Stamens twelve or more. Filaments articulated. Fertile flower solitary, stipitate, naked. Styles three, bifid. Capsule three seeded.*—The species *Ipecacuanha is procumbent, with opposite, obovate, oblong or linear leaves; peduncles axillary, one flowered, elongated.*

The genus was placed by Linnæus in the class *Dodecandria*, order *Trigynia*. Michaux, considering as separate male flowers, the bodies of stamens which correspond, in number, to the nectaries or lacinulæ, has referred the genus to *Monœcia*, *Monadelphia*. In this he has been followed by various American botanists.—In natural arrangements this genus is among the *Tricoccæ* of Linn. and *Euphorbiæ* of Juss.

The Euphorbia ipecacuanha is a low, tufted plant, growing in sandy soils in the Middle and Southern states. Michaux remarks, that the plants are sometimes buried in the sand. It is a polymorphous vegetable both in its shape and colour, the leaves continually differing in their outline, even in contiguous plants; and the colour varying from green to crimson.

The root is irregular and fleshy, very large in proportion to the plant it bears, running deep into the sand, sometimes, as Mr. Pursh informs us, extending to the depth of six feet. The stems, from one root, are numerous, erect or procumbent, forming large bunches on the surface of the ground. They are smooth, regularly dichotomous, and jointed at the forks. The leaves are inserted at the joints, opposite, sessile, smooth, having most frequently an oblong shape though different plants possess every intermediate variety in the form of the leaf, from circular to linear. Their size and colour are likewise variable. The flowers are solitary on long peduncles from the forks of the stem. Calyx spreading, divided into five obtuse segments. Inner segments or nectaries five, small, gibbous. Stamens numerous, in five parcels, appearing, at different times, two or three together, with double anthers. The fertile flowers have a large, roundish, drooping, pedicelled germ, crowned with six revolute stigmas. Capsule three celled.

The dried root of the Euphorbia ipecacuanha is of a greyish colour outside, and white within. It is light and brittle and has about the hardness of cork. To the taste it is sweetish and not particularly unpleasant.

I subjected some portions of the root to chemical examination and obtained the following results.—Sulphuric ether digested on the powdered root dissolves a part of it; and this ethereal solution gives a precipitate, if alcohol is added to it.—Alcohol alone takes up another portion of the root, and assumes a pearly turbidness after water is added. Both the ethereal and alcoholic solutions, evaporated to dryness, leave a residuum which is fusible and inflammable. The decoction gives no precipitate with gelatin or sulphate of iron. With alcohol it gave out a white precipitate which rendered the solution turbid, and subsided in flocks. The cold infusion exhibited the same phenomena in a smaller degree. From these appearances we may infer that the root contains caoutchouc, resin, mucus and probably fæcula.

The Euphorbia ipecacuanha has long been known to possess the same property, which is so frequent in its genus, of exciting the stomach powerfully as an emetic. The appropriation of its specific name seems even to imply that such a property had been recognised in this species in a more eminent degree, than in the rest. It does not appear, however, that it has ever continued long in use, this being prevented, probably, by the suspicious character of the race of plants to which it belongs. The late Dr. Barton mentions this vegetable among his indigenous emetics, but considers it too violent and uncertain to be depended on as a safe medicine.

Within a few years the plant has been attended to by some medical gentlemen in Philadelphia, who report more favourably of its powers and mode of operation; and consider it as a safe, certain and manageable emetic, applicable to most of the cases in which medicines of this kind are called for.

Being desirous to obtain personal knowledge of the medicinal character of this vegetable, I instituted trials with different parcels of the dried root, some of which were gathered by myself, in flower, near Philadelphia, and the rest sent me by friends from Baltimore and Washington. Portions of these roots were given to a variety of patients in the Dispensary and Almshouse by myself and by other physicians, who have obliged me by communicating the results of their observations. These experiments have led to the conclusion that the Euphorbia ipecacuanha in doses of from ten to twenty grains is both an emetic and cathartic; that it is more active than ipecacuanha in proportion to the number of grains administered; that in small doses it operates with as much ease as most emetics, in a majority of instances. If it fails, however, at first, it is not so safely repeated as the other emetics in common use. Given in large doses it excites active and long continued vomiting, attended with a sense of heat, vertigo, indistinct vision, and prostration of strength. I have not ventured upon any large dose myself, but have been informed, that such is the effect, by those who have given the root in doses of two scruples and upwards. The plant appears to differ from the South American Ipecacuanha in having the degree of its operation proportionate to the quantity taken; the process of vomiting not being checked by the powder being thrown off of the stomach, as frequently happens, when common Ipecac is given in large doses.

At my request, Dr. James McKeen made this plant and another species, *E. corollata*, the subjects of an inaugural dissertation at Harvard University, in 1820. As his observations have been made with some care, and illustrate very fairly the action of the medicine, I insert the principal cases from his manuscript.

"*Case* I. The first experiment," he observes, "made with this species of the Euphorbia was upon a man of intemperate habits, about twenty seven years of age, and who appeared to be a candidate for Delirium Tremens. I gave him ten grains. He told me that it always required powerful doses of medicine to produce any effects upon his stomach or bowels, but as I was then a stranger to the powers of the Euphorbia ipecacuanha, it was thought prudent not to hazard a large quantity until something had been ascertained of its strength. When I called in the morning after it was taken, I learned that the medicine had produced a gentle purging, preceded by a considerable degree of nausea, but that there had been no vomiting.

"*Case* II. The next fair opportunity which occurred for experiment was in the case of a female about thirty seven years of age. This woman, for a considerable portion of her life, had suffered from syphilis; nothing remained now, however, specifically of this kind, excepting the marked effects of a constitution shattered by disease. I gave her at first ten grains of the Euphorbia ipecacuanha, and in twenty minutes, no signs of vomiting occurring, I gave her eight grains more, and kept adding to the quantity, which she had taken, until it amounted in the whole to forty grains. I remained by this patient until vomiting commenced, which was precisely thirty five minutes after the exhibition of the first ten grains. As the influence of the mind, in contemplating the effects of an emetic, will often induce its more speedy operation, I diverted the patient's attention as much as possible, that no consequences might ensue, but such as were produced by the specific action of the medicine. As soon as I ascertained that this Euphorbia ipecacuanha was likely to produce effectual vomiting, I left the house. About thirty hours afterwards I called to see this patient, and with much surprise found that the quantity I had given her had continued to operate by emesis and catharsis ever since. She was, however, very little exhausted, and there was nothing of cramp either on the stomach or extremities which so often distress those who are too severely vomited. About this time there was a cessation of vomiting without the assistance of remedies. Two days afterwards this woman told me she had not been as well as she then was for a number of years. The powerful vomiting produced a considerable degree of dizziness, but this went off in the course of twenty four hours. I had quite despaired of vomiting this patient with the Euphorbia ipecacuanha. In no instance afterwards was this medicine more than half as long in producing vomiting as it was in this case.

Case III. A girl of about eighteen years of age, whose manner of living was similar to that of the person mentioned in the preceding case, applied to me for an emetic; I gave her thirty grains of the Euphorbia ipecacuanha, and told her to take half of this quantity, and if it did not operate in half an hour, she might take the remainder. Contrary to my injunctions she took the whole at a single draught. In fifteen minutes her attendants told me she began to vomit, and continued to throw up, at intervals, smartly for five hours, and was purged seven or eight hours more. For some time after this she complained much of dizziness.

Case IV. As I had found, in the first trial, that ten grains of the Euphorbia ipecacuanha failed to produce vomiting, I tried the same dose upon another subject, which was a woman of about forty eight years of age, to determine if so small a quantity would produce vomiting. In about fifteen or twenty minutes after the medicine was received into the stomach, it began to operate. After she had vomited three times, it commenced purging, and produced three or four evacuations. This woman did not complain of any dizziness, as those did in the two preceding cases.

Case V. A woman about thirty one years of age took fifteen grains of the Euphorbia ipecacuanha; in seventeen minutes it began to operate, and vomited the patient every few minutes, until the operation amounted to five or six times, and afterwards a moderate purging ensued. The operation, in this case, was more satisfactory than any preceding ones, as it effectually evacuated the stomach and bowels, without a too long continuance of the vomiting.

Case VI. A man of forty years of age was seized with symptoms of fever. Four grains of sulphate of copper and twelve of common ipecac were given. This dose produced no emetic operation, but occasioned violent purging. Forty eight hours after, I gave him twenty grains of Euphorbia ipecacuanha, in powder, which produced very effectual evacuations from the stomach, vomiting him eight or nine times; after which he had one or two alvine discharges.

Case VII. In one instance, for experiment, I gave four grains of this plant; but it neither affected the stomach nor bowels, nor the feelings of the patient, nor his pulse."

From what is now known respecting the Euphorbia ipecacuanha, we are justified in considering it an active emetic, and, if prudently administered, more safe than a majority of the species of its genus. It wants, however, the peculiar mildness of the officinal Ipecacuanha, which, in cases of slow operation, permits the dose to be accumulated by repetition, until its due effect takes place, without danger of excessive violence in the length and degree of evacuation, and without an injurious impression on the nervous system. This, indeed, appears to be the common defect of the active North American emetics hitherto examined. And until a more extensive examination has brought to light new substances of this class, or better defined the modes of preparation and use of those already known, we cannot wish that the South American drug should be diminished in our markets, or less familiar to our physicians.

BOTANICAL REFERENCES.

Euphorbia Ipecacuanha, LINN. *Sp. pl.*—WILLD. ii. 900.—MICHAUX, *Flora.* ii. 212.—PURSH, ii. 606.—*Botanical Magazine, t.* 1494.—Euphorbia inermis, &c.—GRONOVIUS, *Virg.* 74.—Tithymalus flore minimo herbaceo?—CLAYTON, *Phil. trans. abr.* viii. 331.

MEDICAL REFERENCES.

SCHŒPF, *Mat. Med.* 74.—B. S. BARTON, *Coll.* 26.—W. P. C. BARTON, *Veg. Mat. Med.* vol. i.

PLATE LII.

Fig. 1. *Euphorbia Ipecacuanha.*

Fig. 2, 3, 4, 5. *Different forms and sizes of the leaf observed in individuals of this species.*

Fig. 6. *Red variety of the leaves.*

Fig. 7. *Calyx.*

Fig. 8. *Calyx opened, with five of the stamens expanded.*

Fig. 9. *A perfect flower.*

Fig. 10. *Styles and stigmas magnified.*

PL. LIII.

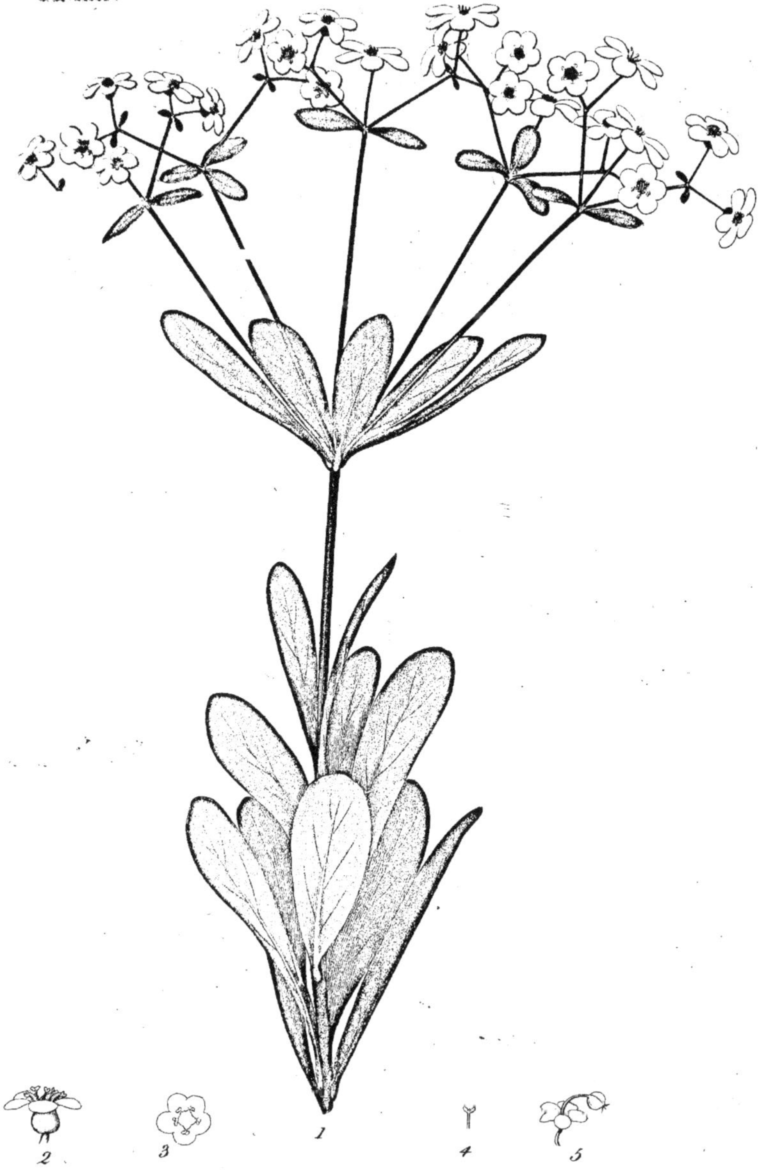

Euphorbia corollata

Annin & Smith.

EUPHORBIA COROLLATA.

Large flowering spurge.

PLATE LIII.

In point of stature and the shewy appearance of its flowers, this species of Euphorbia differs eminently from that described in the last article. In the common features, however, of the genus, such as its lactescence, its taste, and its medicinal powers; the consanguinity of the two plants evidently appears. I am not aware that this species has been much known for its operative qualities, until within a very recent period. The indians were, indeed, acquainted with the medicinal properties of more than one species of Euphorbia. They doubtless made use of the E. ipecacuanha, and not impossibly of the present species also. In Mr. Clayton's letter to Dr. Grew, contained in the Transactions of the Royal society for 1730, and which we have noticed in speaking of Aletris farinosa, the writer states, that the Aborigines made use of "the roots of Tythymal, of which there are two sorts, the one *flore minimo herbaceo*, the other *flore albo*. The flower of this last," he says, "is small, but large in comparison with the other. They are *repentes*, and grow in old manured grounds. They chiefly make use of the latter of these, and it is a most excellent purge, though it sometimes vomits. It is quick but moderate in its effect, and has this peculiarity, that it opens the body, when other more violent purgatives will not move it." We might safely conclude that the white flowering species, here noticed, is the Euphorbia corollata, were it not for the term *repentes* applied to both plants. It is not improbable that in this respect, the writer might have been misinformed.

Pursh informs us that Euphorbia corollata grows in dry fields from Canada to Carolina. I have never met with it north of Pennsylvania. The drawing which illustrates our description is from a specimen cultivated in the Botanic garden at Cambridge. It is a tall, erect plant, from one to five feet in height, resembling, at a distance, some of the white flowering *corymbiferæ*. It begins to flower in June, but is not fully expanded until July or August. Its specific character is as follows. *Umbel five rayed, three parted, dichotomous; leaves and involucra oblong, obtuse; segments of the calyx obovate, petaloid, coloured.* The shape of the leaves is subject to variety, as is also their smoothness or hairiness.

This plant has a large branching root which sends up a number of stems, frequently from two to five feet in height. They are erect, round and in most instances simple. The leaves are scattered, sessile; oblong, obovate or linear, a little revolute at the margin, smooth in some plants, very hairy in others. The stem divides at top into a large five rayed umbel, supported by an involucrum of as many leaves. Not unfrequently a small axillary branch or two arise from the sides of the stem below the umbel. The rays of the umbel are repeatedly trifid or dichotomous, each fork being attended by two leafets and a flower. The top of the stem or centre of the umbel is turgid, and often bears a precocious flower. The calyx is large, rotate, white, with five obtuse petal-like segments, from which the name of the species has been taken. The nectaries or inner segments are five, very small, obtuse projections situated at the base of the segments. Stamens a dozen or more emerging two or three at a time, with double anthers.

Germ pedicelled. Capsule three celled. A great portion of the plants are wholly stamini-ferous.

The results of a short chemical examination of this plant were very similar to those afforded by E. ipecacuanha. The ethereal solution was made turbed by alcohol, and the alcoholic by water. The precipitate in the last instance seemed denser and more abundant than it was in the former species. The decoction deposited a mucus or feculent substance, by means of alco-hol, as in the other plant. The same sweetish taste characterised the solutions of both veg-etables.

It has been observed, by late experimenters in vegetable chemistry, that most of the lactes-cent or milky plants contain caoutchouc. That they contain a substance of this nature, which is dissolved by ether and not by alcohol, I am able to attest from the examination of various lactes-cent plants inserted in this work, and some others.

The properties of Euphorbia corollata have been lately brought into notice by W. Zollic-koffer, M. D. of Baltimore, to whom I was first indebted for my specimens of the root and living plant; and who has furnished me with a variety of facts relating to its properties. Dr. Z. states that this plant is quite common in some parts of the state of Virginia. In some districts of Maryland, and more particularly in Anne Arundel county, it grows in the greatest abundance, where it is recognised by the common appellations of Milkweed, Snake's milk, Ipecacuanha and Indian Physic. It delights in a poor, dry, and sandy soil. It is seldom or never found growing in the woods, but in fields that are cultivated every two or three years. The farmers have frequently told him that it is very hurtful to small grain, when it grows in great quantities, and the common means that are made use of, such as ploughing and harrowing, in order to kill *bluegrass*, have the effect of increasing the quantity and rapid growth of this plant. It is never eaten by animals. The root is sometimes used as an emetic by the country people; and it is esteemed in the cure of dropsy. The stalks, which arise from the common trunk of the root, are sometimes as many as thirty, and from this down to a single one. The largest roots, which he recollects seeing, measured from an inch to two inches and a half in circumference. He has been in the habit of using the Euphorbia corollata, for some time past in practice, as an emetic, in the place of the Ipecacuanha of the shops; and thinks it in no respect inferior to this article. Combined with opium and the Sulph. potassæ in the proportions of the Pulv. doveri, he has found it to be a valuable diaphoretic. The contused root, in its recent state, will excite inflammation and vesication, when applied to any part of the body; which generally goes off in the course of four or five days whithout being attended with any inconvenience whatever. He was led to give it a number of trials in this way from the circumstance of his face becoming considerably inflamed after having handled large quantities of the root. As an expectorant, this plant, he says, is deserving of the attention of practitioners.

Dr. Z. has furnished me with minutes of seventeen cases, in which he administered the powdered root of this plant in doses of from ten to twenty grains. In all of these it operated by vomiting, with the exception of two cases, where it produced nausea, followed by catharsis. Having tried a variety of preparations, he states, that the extract may be given in doses of from five to eight grains; the wine prepared in the same way as Vinum ipecacuanhæ, in dose of an ounce or an ounce and an half. Of the root in powder from fifteen to twenty grains was found a proper emetic. He considers this medicine as having a peculiar advantage in possessing no unpleasant taste; being only followed by a sense of heat in a few minutes after it is taken. But this is by no means lasting, nor does it produce any material uneasiness.

In some experiments, to determine the soluble portions of this root, Dr. Z. found that two thousand one hundred and sixty grains of the recent root afforded one hundred and two grains of watery extract; and a like quantity by digestion in alcohol gave one hundred and twenty three grains of alcoholic extract. He did not observe any difference in the activity of these two extracts.

Dr. McKeen, whose Dissertation on the species of Euphorbia has been already cited in the last article, has detailed the circumstances of twelve cases, in which he administered the Euphorbia corollata. His experiments differ from those of Dr. Zollickoffer, in the quantity of the root used, being always smaller. The doses, which he gave, were from three to twelve grains of the powder. In every instance the medicine operated as a cathartic. In most of the cases nausea was produced, but in three only, out of the whole number, it was followed by vomiting.

In one case a dose of three grains proved actively cathartic in four hours. In another five grains produced vomiting. In a third no effect was experienced from twelve grains, except that of a moderate laxative. In one instance twenty grains were given, which produced vomiting three times, followed by about twenty alvine evacuations.

I have placed portions of this plant in the hands of several practitioners and medical students, with a request to be informed of the effect, when suitable opportunities for its exhibition had occurred. In a majority of the instances I have been told, that a cathartic operation had followed its use; and sometimes, though less frequently, an emetic. It rarely has proved inactive.

The Euphorbia corollata must undoubtedly be ranked among the more efficient medicines of the evacuating class. Dr. McKeen concludes, from his experiments, that it is a very certain purgative, possessing, he thinks, about double the strength of jalap. It exerts its cathartic efficacy in doses of less than ten grains. If given to the amount of fifteen or twenty grains, it is very sure to prove emetic; the proportion of its failures, being not greater than occurs in the use of other emetic medicines. The only inconveniences which have come to my knowledge, as attending it, are, that if given in small doses, for a purgative, it is apt to produce nausea; while in the large doses suitable for an emetic, it sometimes has induced a degree of hypercatharsis. But it must be observed, that many of the medicines, in common use, may occasion similar consequences in persons of peculiar habit and irritable fibre. Future experiment will, no doubt, determine whether the Euphorbia corollata is any more irregular and unmanageable than other medicines of its kind, or whether it is entitled to a permanent and useful place in the Materia Medica.

Many, and perhaps all the species of Euphorbia are powerful external stimulants. Several are used as a sort of caustic to destroy warts. The gum, called Euphorbium, produced by the Euphorbia officinarum, is a strong vesicatory, employed by farriers, and sometimes used to adulterate the plaister of Cantharides. The blistering power of E. corollata has been stated by Dr. Zollickoffer. This active genus of plants deserves a thorough investigation with a view to this particular property, to determine whether they are safe and manageable vesications, or virulent and uncertain.

BOTANICAL REFERENCES.

Euphorbia corollata, Linn.—Willd. ii. 916.—Michaux, ii. 210.—Pursh, ii. 607.—Tithymalus marianus, &c.—Plukenet, *Mant.* 182. *t.* 446. *f.* 2.

MEDICAL REFERENCES.

Clayton, *Philosophical transactions abridged,* viii. 331.—Zollickoffer, *Materia Medica. Baltimore,* 1819.

PLATE LIII.

Fig. 1. *Euphorbia corollata, the top of a plant rather below the common size.*

Fig. 2. *Barren flower.*

Fig. 3. *Calyx not fully expanded.*

Fig. 4. *Stamen.*

Fig. 5. *Fertile flower.*

PL. LIV.

Polygala rubella.

Annin & Smith.

POLYGALA RUBELLA.

Bitter Polygala.

PLATE LIV.

THIS plant is interesting from the curious manner in which a part of the fruit is produced, by a kind of imperfect flower growing close to, and in some instances under, the surface of the ground. It is not the only species of the Polygala which has this peculiarity. I have often observed little shoots at the root of P. paucifolia, one of the most beautiful of the genus, bearing apterous flowers and subterranean fruit, precisely like those represented in our plate. The P. polygama, of Walter and Pursh, if, indeed, it is a distinct species, has the same remarkable mode of growth. It is difficult to imagine what end is attained by nature in this singular arrangement, by which a part of the seeds are ripened in the sun, while the rest, like the fruit of Arachis hypogæa, is buried from the light. To the eye there is no difference between seeds taken from the upper or lower racemes of the plant. It would be worth while to ascertain if the two will vegetate equally well.

The genus is marked by a *calyx of five leaves, two of which are wing-like and coloured. Capsule obcordate, two celled and two valved.* The species *rubella* has its *stems simple; leaves linear-oblong, mucronated; flowers racemed, those of the stem winged, those of the root apterous.*

Class *Diadelphia*, order *Octandria;* natural orders *Lomentaceæ*, Linn. *Pediculares*, Juss.

The Polygala rubella, here described, is the plant designated by that name in Muhlenberg's catalogue, as I have formerly learnt from the author himself. There is little doubt that Willdenow's plant is the same, described from an imperfect specimen. It is found in dry, sandy, or gravelly soils in many parts of the United States, and flowers in June and July.

Root somewhat fusiform, perennial, branching. Stems numerous, ascending, smooth, angular, simple. Leaves scattered, smooth, the lower ones obovate, smaller; the upper ones linear-lanceolate, obtuse, mucronated, sessile. Flowers purple, short-crested, in terminal racemes.

Bractes small, ovate-lanceolate, caducous. Wings of the calyx rhomboid-oval, obtuse, with a slight middle nerve. Corolla small, closed, of three segments, the middle one largest and crested by the division of its sides and extremity. Anthers eight, forming a double row, the filaments coalescing. Germ compressed, inversely heart-shaped; style deflexed; stigma bearded inside, with a prominence below it. Capsule inversely heart-shaped, nearly smooth, margined, and invested with the wings of the calyx. Seeds two, obovate, hairy, with a transparent appendage or strophiole on the inside. From the base of the stems proceed a number of prostrate shoots situated upon, and sometimes nearly under the ground, bearing a row of incomplete fertile flowers. These flowers are furnished with a calyx without wings, a minute corolla and stamens, and a short style. The germ and fruit precisely resemble those of the more perfect flowers.

Like some of the European species which it resembles in habit, this plant is a strong and permanent bitter, imparting its sensible properties both to spirit and to water. I digested a portion of the dried plant in ether, and added alcohol to the solution. No change was visible at the time of mixture, but on standing till the ether had partly evaporated, the alcohol became turbid. A tincture of the plant was not immediately affected by adding water, but on standing over night it became very turbid, and in a few days deposited a large precipitate. The bitterness, which is probably of the extractive kind, was communicated to cold, as well as hot water; and to alcohol. The aqueous solutions appear strong enough to represent the virtues of this vegetable.

The Polygala rubella, from its extreme bitterness, has attracted the notice of various medical practitioners in the Northern states. I have been assured by those who have tried its efficacy, that the infusion administered in small doses, proves a useful tonic and stimulant to the digestive organs. In large doses it opens the body and excites diaphoresis. Its powers appear to resemble those of Polygala vulgaris and P. amara of Europe, to which it has a close botanical resemblance; and which have enjoyed a certain degree of medicinal reputation as tonics and expectorants.

BOTANICAL REFERENCES.

Polygala rubella, MUHLENBERG, *Catal.*—WILLD. iii. 875.—PURSH, ii. 464.—Polygala polygama ?—NUTTALL, *genera*, ii. 87.

PLATE LIV.

Fig. 1. *Polygala rubella.*
Fig. 2. *A flower.*
Fig. 3. *Calyx.*
Fig. 4. *Corolla magnified.*
Fig. 5. *Fruit of ditto.*
Fig. 6. *Body of stamens.*
Fig. 7. *Pistil.*

NYMPHÆA ODORATA.

Sweet scented Water lily.

PLATE LV.

The common Water lily, of North America, very much resembles that of Europe in its external form, but differs remarkably in the fine fragrance of its flowers, those of the old continent being nearly destitute of odour. It belongs to a very beautiful tribe of aquatic plants, a great part of which are natives of the torrid zone. Those species which support the cold of our northern latitudes, are enabled to do so only by the depth of water, under which it is their habit to vegetate. Nature has provided a sort of spontaneous hotbed for these plants, by placing their roots at such a depth from the surface of the element in which they grow, that the frost, which would otherwise prove fatal, does not reach them at the coldest season.

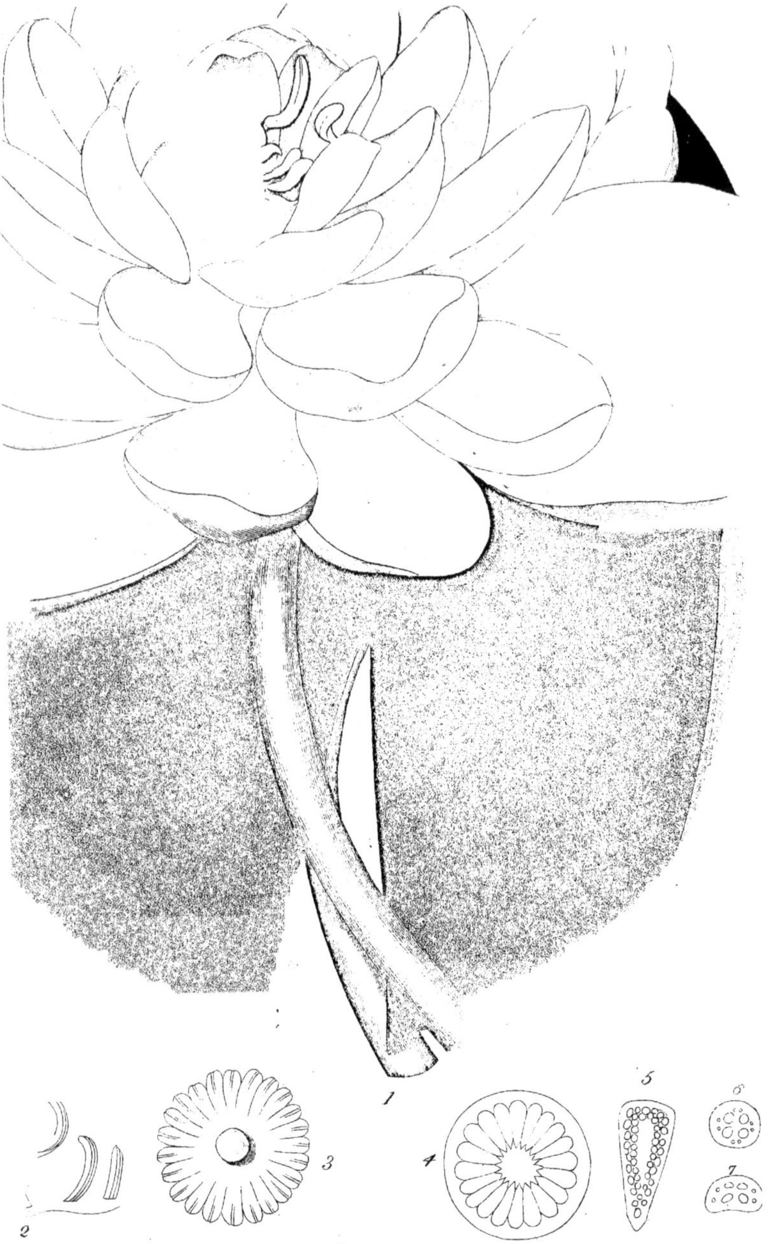

Nymphæa odorata

Annin & Smith

The Nymphæa odorata, the finest of the northern species, grows abundantly in most parts of the United States, about the edges of rivers and ponds, where the water is more than a foot in depth. It is one of the largest of our native flowers, and though it has often been represented as inferior, in size, to the water lily of Europe, I am sure that this comparison can only have resulted from the inspection of cultivated specimens. The annexed drawing was made from a full grown and fully expanded specimen, and is actually smaller than the flower from which it was taken.

Every angler is familiar with the leaves and stems of this plant, which, with a few similar aquatics, forms floating beds about the edges of deep fresh waters, affording to the fish a favourite shelter from the light; and often rendering them more essential service, by entangling the hooks and lines of their pursuers.

The roots of this plant creep through the muddy bottoms of ponds to a great extent. They are very rough, knotted, blackish, and as large as a man's arm. The porous stalks, which proceed from these, are bouyed up by the quantity of air they contain, and continue to be elongated till they reach the surface of the water, which is often at the height of several feet. The upper side of the leaves has a highly repellent power for water, owing to its finely polished surface, from which the fluid rolls off as from a coating of oil. When the buds have attained to maturity, they emerge and expand their flowers. This takes place in the morning; and when the sun is bright, a bed of these flowers presents a truly magnificent spectacle. Owing to the concavity of the calyx and petals they continue to float during a great part of the day. They are seldom elevated from the surface, except when the stem is uncommonly large, or pushed upward by some displacement of the adjacent leaves. At night, or before, the flowers close, and either rest on the surface or sink beneath it till the subsequent day. When flowering is over, the germ sinks to the bottom and there ripens its fruit.

The genus Nymphæa is now separated from some other plants formerly attached to it by the following character. *Calyx four or five leaved; petals many, inserted into the germ below the stamens; stigma radiated, sessile with a tubercle in the middle; berry many celled, many seeded.* This species very nearly resembles the *N. Alba* of Europe, but appears distinct by the following marks. *Leaves orbicular-cordate, entire, the lobes acuminate, and veins prominent beneath; calyx four-leaved, equal to the petals.*—Linnæus placed this genus in his *Miscellaneæ*, and Jussieu with the *Hydrocharides.*

The stalks, both of the leaves and flowers, spring directly from the root. They vary in length from one foot to five or six, according to the depth of the water. The petioles are somewhat semicircular, the scapes round. Both are perforated throughout by long tubes or air-vessels which serve to float them. The leaves, which swim on the surface, are nearly round with a cleft or sinus extending to the centre, at which the petiole is inserted in a peltate manner. The lobes on each side of this sinus are prolonged into an acute point. The upper surface is of a bright glossy green almost without veins; the lower surface is reddish and marked by a multitude of strong prominent veins diverging from the centre. The calyx has four lanceolate leaves, green without and white within. Petals numerous, lanceolate, of a delicate whiteness, with sometimes a tinge of lake on the outside. Stamens numerous, yellow, in several rows; the filaments dilated, especially the outer ones, so as to resemble petals; the anthers in two longitudinal cells growing to the filaments and opening inwardly. The stigma has from twelve to twenty four rays, very much resembling abortive anthers, at first incurved, afterwards spreading. At the centre is a solid hemispherical protuberance, usually called a nectary, but which appears to me more like the true stigma.

The roots of this plant are among the strongest astringents, and we have scarcely any native vegetable which affords more decided evidence of this property. When fresh, if chewed in the mouth, they are extremely styptic and bitter. Their decoction instantly strikes a jet black colour with sulphate of iron, and yields a dense, white precipitate to a solution of gelatin. With alcohol it deposites a slight flocculent substance resembling fæcula. Tannin and gallic acid in large quantities are to be considered its most characteristic ingredients.

The flowers have a delicious odour, hardly surpassed by any perfume which the summer produces. This fragrance is perfect only when the flowers are fresh, and, as they droop, becomes contaminated with the common smell of aquatic plants. It is peculiar in its character, and resembles that of no other plant with which I am acquainted. I have several times attempted to separate this perfume by distillation both with water and spirit, but have never succeeded in preserving it in the faintest degree. It is much more fugacious than the perfume of roses, and seems to be destroyed by the application of heat. Possibly the employment of a large quantity of flowers at a time might yield a better product. The stamens appear more odorous than the petals, or at least preserve their odour longer in drying.

The roots of the water lily are kept by most of our apothecaries, and are much used by the common people in the composition of poultices. They are, no doubt, often injudiciously applied to suppurating tumours, since their astringency must be rather discutient, than promotive of suppuration. They are occasionally used by physicians in cases where astringent applications are called for, and answer a purpose somewhat analogous to that of lead poultices and alum curds. The roots, which, when fresh, are large and fleshy; in drying, lose a great part of their weight and size, becoming spongy and friable.

The Nymphæa alba of Europe, which appears perfectly similar in its qualities to the American plant, was celebrated by the ancients, [*Note* C.] as an antaphrodisiac, and as a remedy in dysentery and some other morbid discharges. To the latter purpose its astringency might, in some instances, make it well suited. The roots and seeds of the Nymphæa lotus were used by the ancient Egyptians as bread.

BOTANICAL REFERENCES.

Nymphæa odorata, WILLD. *Sp. pl.* ii. 1153.—*Bot. Mag.* 819.—*Bot. Repository,* 297.—PURSH, ii. 368.—Nymphæa alba, MICHAUX, i. 311.—WALTER, *Carol.* 155. Castalia pudica, SALISBURY, *Annals of Bot.* ii. 71.

MEDICAL REFERENCE.

CUTLER, *Amer. Transactions,* i. 456.

PLATE LV.

Fig. 1. *Leaf and flower of Nymphæa odorata.*
Fig. 2. *Different stamens from the same flower.*
Fig. 3. *Stigma.*
Fig. 4. *Section of the germ.*
Fig. 5. *A cell of the germ magnified.*
Fig. 6. *Section of the scape.*
Fig. 7. *Section of petiole.*

LVI.

Prinos verticillatus

Annin & Smith.

PRINOS VERTICILLATUS.

Black Alder.

PLATE LVI.

After the leaves have fallen in autumn, this shrub becomes conspicuous by its glossy scarlet berries, which adhere in bunches, for a long time, to the sides of the branches. Of the objects which impart any liveliness to this season of decay, the most noticeable are those which change the hue of their leaves from green to red, as the oaks, the vaccinia, *&c.* those which flower late, as the Hamamelis, and those whose fruit attains to maturity under the influence of frost, and appears fresh and vegetating, while other things are withering about them. The species of Prinos are of the last description.

This genus consists of shrubs, a part of which are deciduous, and a part evergréen; bearing small lateral or axillary flowers. It is nearly related to the *Ilices* or Hollys, differing chiefly in the number of its parts. Its character is formed by *a six cleft calyx, a monopetalous subrotate six cleft corolla, and a six seeded berry.* The Prinos verticillatus has *its leaves deciduous, oval, serrate, acuminate, slightly pubescent beneath; flowers axillary, aggregate.*

These shrubs have usually been referred to *Hexandria Monogynia.* The present species and some others having different flowers on separate plants, Michaux was induced to place them in Diœcia. The natural orders to which they are assigned are *Dumosæ* of Linn. and *Rhamni* of Juss.

The Black Alder, for so the shrub is usually called, is found in swamps and about the edges of streams and ponds from Canada to the Southern states. It is irregular in its growth, but most commonly forms bunches six or eight feet in height. The leaves are alternate or scattered, on short petioles, oval, acute at base, sharply serrate, acuminate, with some hairiness, particularly on the veins underneath. The flowers are small, white, growing in little tufts or imperfect umbels, which are nearly sessile in the axils of the leaves. Calyx small, six cleft, persistent. Corolla monopetalous, spreading, without a tube, the border divided into six obtuse segments. The stamens are erect, with oblong anthers. In the barren flowers they are equal in length to the corolla, in the fertile ones, shorter. The germ, in the fertile flowers, is large, green, roundish, with a short neck or style, terminating in an obtuse stigma. These are followed by irregular bunches of bright scarlet berries, which are roundish, supported by the persistent calyx, and crowned with the stigma, six celled, containing six long seeds, which are convex outwardly and sharp edged within. These berries are bitter and unpleasant to the taste, with a little sweetness and some acrimony.

The bark of the Black alder is moderately bitter, but inferior in this respect to many of our shrubs and trees. It discovers very little astringency either to the taste, or to chemical tests. A decoction which I made of the dried bark underwent no alteration on the addition of dissolved gelatin, and only changed to a dark green with the sulphate of iron. Alcohol produced hardly any change. The tincture, in alcohol, was found moderately bitter, and was not altered by water.

The Black alder has had a considerable reputation as a tonic medicine, perhaps more than it deserves. The late Professor Barton tells us, that the bark has long been a popular remedy in different parts of the United States, being used in intermittents and some other diseases as a substitute for the Peruvian bark; and on some occasions, he thinks it more useful than that article. "It is employed both in substance and in decoction, most commonly, however, in the latter shape. It is supposed to be especially useful in cases of great debility accompanied with fever; as a corroborant in anasarcous and other dropsies, and as a tonic in cases of incipient sphacelus or gangrene. In the last case," he says, "it is unquestionably a medicine of great efficacy. It is both given internally and employed externally as a wash."

Dr. Thacher recommends a decoction or infusion of the bark taken internally in doses of a teacupful, and employed also as a wash, for the cure of cutaneous eruptions, particularly of the herpetic kind.

I have had but little experience with the bark of the Prinos which gave me much satisfaction. Indeed the tests of tonic remedies are of a more ambiguous kind than those of most other medicines. Vegetable barks, which are bitter and astringent, are generally tonic, if they have no more striking operation; and in this property they differ in a degree somewhat proportionate to their bitterness and astringency. Judging by these criterions, the Prinos is not entitled to hold a very exalted rank in the list of tonics. As a bitter it is at best but of the second rate, and in astringency it falls below a multitude of the common forest trees.

The berries are recommended by the writers above cited, as possessing the same tonic properties with the bark. They certainly possess some activity, which, in large quantity, is not of the tonic kind. I have known sickness and vomiting produced in a person by eating a number of these berries found in the woods in autumn.

BOTANICAL REFERENCES.

Prinos verticillatus, Linn, *Sp. pl.*—Pursh, i. 220.—Prinos Gronovii, Michaux, ii. 236.—Prinos padifolius, Willd. *Enum. Berol.* 394.

MEDICAL REFERENCES.

B. S. Barton, *Collections*, ii. 5.—Thacher, *Disp.* 324.

PLATE LVI.

Fig. 1. *Prinos verticillatus, a branch in flower.*
Fig. 2. *Ripe berries.*
Fig. 3. *Calyx magnified.*
Fig. 4. *The rest of the flower ditto.*
Fig. 5. *Stamen of the barren flower magnified.*
Fig. 6. *Germ of the fertile flower ditto.*

PL. LVII.

Sabbatia angularis

Annin & Smith.

SABBATIA ANGULARIS.

American Centaury.

PLATE LVII.

UNDER the name of *Chironia angularis*, this plant has been familiar to physicians in the United States as a native bitter. As it wants the most distinguishing characters of Chironia, while it has others of a very different kind, particularly in the anthers and stigma; I have followed the example of Pursh and others in referring it to the genus Sabbatia of Adanson.

This genus is characterised by *a persistent calyx from five to twelve parted; a corolla from five to twelve parted; anthers finally revolute; stigma two parted, spiral; capsule one celled.* The Sabbatia angularis differs from the rest of the genus in being *erect, the leaves clasping, peduncles elongated and corymbose, segments of the calyx lanceolate, half as long as the corolla; stem square and winged.*

Class *Pentandria*, order *Monogynia;* natural orders *Rosaceæ*, Linn. *Gentianæ*, Juss.

This plant grows in damp rich soils throughout the Middle and Southern states, and is most commonly known by the name of *Centaury.* It is commonly from one to two feet high. The stem is erect, smooth, square, with the angles winged. Branches axillary, opposite. The leaves are opposite and ovate, but vary in length and width. They are heart-shaped at base, clasping half the stem, nerved, smooth, entire, acute. Flowers terminal, forming a large corymb. Tube of the calyx angular, with five broad segments. Corolla five parted with oval segments twice as long as the calyx. The anthers are oblong and slightly recurved at the time when the flower first opens. After shedding their pollen they become revolute and curl up, but never assume the spiral form like the anthers of Chironia. Germ ovate; style longer than the stamens, declined; stigma two parted, the segments separate at first, but gradually becoming twisted spirally together. Capsule one celled, two valved.

Every part of this plant is a pure and very strong bitter. In this quality, as well as in its medicinal properties, it is resembled by several other species of the same genus. An extractive principle appears to be the seat of this property, as it is communicated alike to alcohol and water, and as the solutions in these fluids do not occasion precipitates from each other. There appears to be no astringency in the vegetable.

In the collections for an American Materia Medica by the late Professor Barton, we are told that this plant is a valuable tonic bitter resembling the Centaury of Europe, for which it was used by some practitioners on the supposition of its being the same plant. It had long been a popular remedy, and was much employed in the yellow fever of Philadelphia, in 1793.

In Mr. Elliott's Botany of the Southern states, we are told that the plant, in South Carolina, is a common remedy in intermittent fever. Some of the other species of the same family, particularly S. gracilis, are equally efficacious. It is deserving of remark, that a great number of vegetables, belonging to the same natural order, are highly bitter, and approved as tonic remedies.

From the use I have made of the Sabbatia, I have no hesitation in attesting its utility. It seems to me to rank among the more pure or simple bitters, and acts usefully as a stomachic and promoter of appetite and digestion. Beyond this, I have no experience with it. It may be given in substance or in infusion, but the latter mode is generally preferred. This form is one in which it appears to be largely used by physicians in the Middle states. Dr. Chapman tells us, it is resorted to extensively by every class of practitioners, regular and irregular, in the intermittent and remittent fevers. He thinks it has the advantage over Peruvian bark of being susceptible of employment in every stage of these diseases.

BOTANICAL REFERENCES.

Sabbatia angularis, PURSH, *Flora Amer.* i. 137.—ELLIOTT, *Flora* i. 285.—Chironia angularis, WILLD. *Sp. pl.* i, 1067.—MICHAUX, *Flora*, i. 146.

MEDICAL REFERENCES.

B. S. BARTON, *Coll.* i. 15.—CHAPMAN, *Therapeutics*, ii. 417.—ELLIOTT, *l. c. supra.*

PLATE LVII.

Fig. 1. *Sabbatia angularis.*
Fig. 2. *The stamen before it bursts, magnified.*
Fig. 3. *Stamen after bursting,* - *do.*
Fig. 4. *Pistil magnified with the stigmas not yet twisted.*
Fig. 5. *Do.* *the stigmas having become spiral.*

PL. LVIII

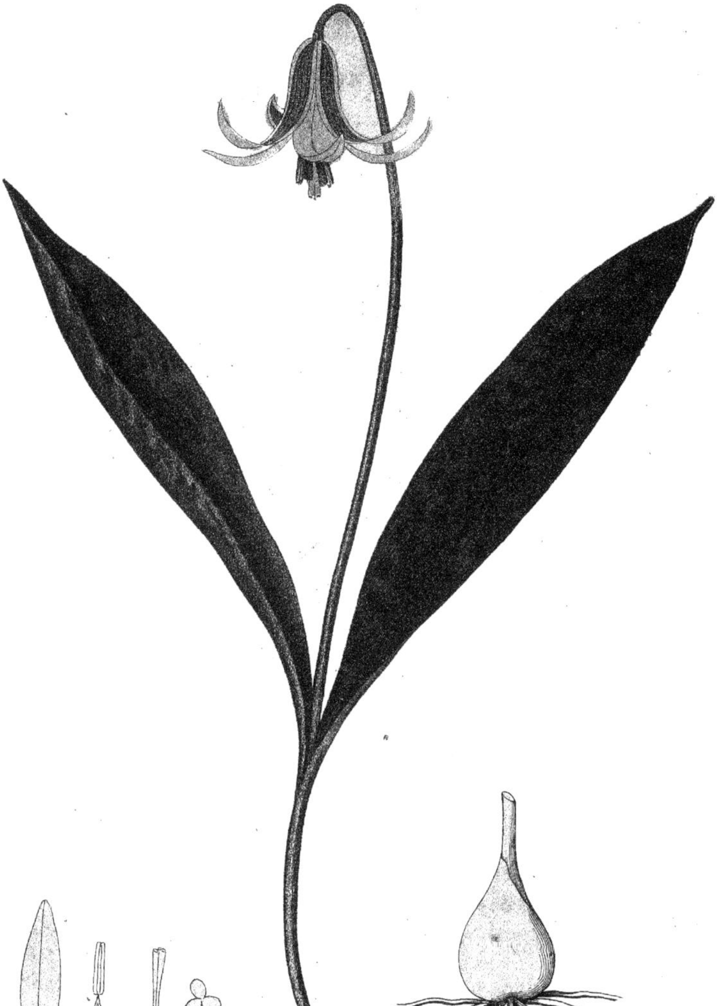

Erythronium Americanum

ERYTHRONIUM AMERICANUM.

Common Erythronium.

PLATE LVIII.

For a considerable time the genus Erythronium was considered as containing only one species, the *E. dens canis* of Europe and Asia. The American plant was considered, by Michaux, as a variety of the European, differing only in colour. Later botanists have, with propriety, separated it, and besides this, one or two other American species have been added to the genus.*

The natural order, called *Liliaceæ* by Linnæus, and *Lilia* by Jussieu, is perhaps not exceeded by any other, in the uniform elegance of all its spe-

* My friend Mr. F. Boott discovered a new species of Erythronium on the Camel's rump mountain in Vermont, which he calls *E. bracteatum.* Its character is *E. foliis inæqualibus, scapo bracteato.* In all the specimens gathered by that gentleman, the leaves were very unequal, one being twice the size of the other; the scape had also a lanceolate bracte near the top. The flower was yellow and about half the size of *E. Americanum.*

cies. The Lily, Tulip, Crown imperial, and Gloriosa are specimens of this order. They belong to the same artificial class and order *Hexandria trigynia*, and have a close affinity in all the parts of their structure. The Erythronium, which is generally called, I know not for what reason, *Dog's tooth violet*, is one of the smallest of the order.

This genus has *no calyx. Its corolla is inferior, six petalled; the three inner petals with a callous prominence on each edge near the base.* The common American plant has its *scape naked*, its *leaves lanceolate and involute at the point; and its style club-shaped and undivided.* It is an early flowering plant, being in blossom in the first part of May. It grows in woods and fields in the Northern and Middle states.

The root is a solid bulb, situated deep in the ground, brown outside, and white and homogeneous within. The whole plant is smooth and glossy. Scape naked, slender. Leaves two, nearly equal, lanceolate, veinless, of a dark brownish green, clouded with irregular spots, sheathing the scape with their base, and terminating in an obtuse callous point. Flower solitary, drooping. Petals six, lanceolate, yellow, the three outermost partly crimson on the outside, the three innermost having an obscure tooth on each side near the base. In a clear sun the petals are expanded and revolute, but at night and on cloudy days, they are nearly closed. Filaments flat, anthers oblong-linear. Germ obovate, style longer than the stamens, club-shaped, three lobed at top and terminating in three distinct, but not detached, stigmas. Capsule oblong-obovate, somewhat pedicelled.

The bulb of this plant, judging from its texture and taste, is almost wholly farinaceous. When dry, it is mealy and free from any unpleasant flavour. Having lost my specimens of the root at the time of preparing this article, I was unable to submit this part to chemical examination. A tincture was prepared from some dried leaves and flowers, which gave evidence of resin being present, when tested with alcohol. Water distilled from the same parts had a rather disagreeable odour.

This vegetable possesses the power of acting on the stomach as an emetic. About twenty five grains of the green root and forty of the recently dried root have produced nausea and vomiting. When the root is fully and thoroughly dried, or when it has been exposed to heat, it appears to lose this property in a great measure.

In its power of acting on the alimentary canal, it resembles many other plants, which are related to it in botanical habit. The Squill, Colchicum, and Aloe are examples of this class, and even the common Daffodil and Tulip are found to be emetic. I have known a family of children to be taken with violent vomiting from having, by mistake, dug up, roasted and eaten some Tulip roots, supposing them to be Artichokes.

It is probable that the medicinal activity of the Erythronium is of a volatile nature, capable of being dissipated by heat. Its farinaceous portion, when duly separated, is no doubt innoxious. Gmelin, in his Flora Sibirica, states, that the Tartars collect and dry the roots of Erythronium *dens canis*, and boil them either with milk or broth, and consider them as very nutritious food. They are said nearly to resemble salep. It is remarkable that farinaceous roots, which possess active and even virulent qualities, do not impart them to the fæcula, which constitutes so large a portion of their bulk. The different species of Arum, Calla, and the Jatropha Manihot are examples of this fact, affording nutritious bread, although their crude juices are more or less poisonous.

The leaves of the American Erythronium are said to be more active than the root, but on this subject I am not fully informed. It is probable that the recent leaves have more activity than the dry.

BOTANICAL REFERENCES.

Erythronium Americanum, KER, *Bot. Mag. t.* 1113.—NUTTALL, *Genera,* i. 223.—E. lanceolatum, PURSH, i. 230.—E. longifolium, POIRET, *Encycl. Methodique.*—E. flavum, SMITH, *Rees' Cycl.*—E. dens canis, MICHAUX, *Flora,* i. 198.

PLATE LVIII.

Fig. 1. *Erythronium Americanum, the flower rather more drooping than common.*

Fig. 2. *One of the inner petals.*

Fig. 3. *Stamen.*

Fig. 4. *Pistil.*

Fig. 5. *Stigma magnified.*

Fig. 6. *Root.*

XANTHOXYLUM FRAXINEUM.

Prickly Ash.

PLATE LIX.

The Prickly Ash is a shrub of middling height, found in woods and moist or shady declivities in the Northern, Middle and Western states. It is rare in Massachusetts and the states north of it, its localities being very circumscribed. After I had taken pains to procure specimens from Connecticut, I accidentally discovered a thicket of the shrubs in a wood in Medford, six miles from Boston.

Late botanists have placed the genus Xanthoxylum in *Pentandria Pentagynia*, although it is diœcious, or rather polygamous. Its *calyx is inferior, five parted; corolla none; capsules from three to five, one seeded.* The *X. fraxineum* is *prickly, the leaves pinnate; leafets ovate, subentire, sessile, equal at base; umbels axillary.*

PL. LIX.

Linnæus placed the Xanthoxyla in his natural order *Dumosæ*, but Smith thinks them better arranged with the *Hederaceæ*. Jussieu places them with his *Terebintaceis affinia.*

The branches of the Prickly ash are covered with strong, sharp prickles, arranged without order, though most frequently in pairs at the insertion of the young branches. Leaves pinnate, the common petiole sometimes unarmed and sometimes prickly on the back. Leafets about five with an odd one, nearly sessile, ovate, acute, with slight vesicular serratures, somewhat downy underneath. The flowers appear in April and May before the leaves are expanded. They grow in sessile umbels about the origin of the young branches, are small and greenish. I have observed them of three kinds, making the shrub strictly polygamous. In the staminiferous flower the calyx is five leaved, the leaves oblong, obtuse, erect. Stamens five with subulate filaments and sagittate four celled anthers. In the place of pistils are three or four roundish corpuscles supported on pedicels from a common base. The perfect flowers, growing on the same plant, have the calyx and stamens like the last; the germs are three or four, pedicelled, and having erect, converging styles nearly as long as the stamens. The pistilliferous flowers grow on a separate shrub. Calyx smaller and more compressed. Germs about five, pedicelled; styles converging into close contact at top, and a little twisted. Stigmas obtuse. All the flowers are destitute of corolla. Each fertile flower produces an umbel of as many stipitate capsules as there were germs in the flower. These capsules are oval, covered with excavated dots, varying from green to red, two valved, one seeded; the seed oval, blackish.

The bark of the Prickly ash has a slight aromatic flavour, combined with a strong pungency, which is rather slow in manifesting itself in the mouth. The leaves are more aromatic, very much resembling, in smell, the leaves of the Lemon tree. The rind of the capsule is highly fragrant, imparting to the fingers, when rubbed between them, an odour much like the oil of lemons. The odorous portion is an essential oil residing in transparent vesicular points on the surface of the capsules and about the margins of the leaves. The acrimony, which resides in the bark, has its foundation in a different principle; being separated by decoction, but not by distillation; at least none of it came over in my experiments, which were repeated with both the green and dried bark. The water in which the bark is boiled has a peculiar pungent heat, which is not perceived when the liquid is first taken into the mouth, but gradually developes itself by a burning sensation on the tongue and fauces. It retains this acrimony after standing a week and more. The leaves do not appear to possess the pungency of the bark, and impart no acrimony to the water in which they are boiled. They abound in mucilage, which coagulates in large films when alcohol is added to the decoction. They seem to possess more astringency than the bark, and strike a black colour with sulphate of iron, while solutions, made from the bark, are but moderately changed by the same test. The alcoholic tincture of the bark is bitter and very acrid. Its transparency is diminished by adding water, and after standing some time it becomes very turbid. Whether the acrimony of this shrub resides in a peculiar acrid principle, or whether it belongs to the resin and becomes miscible with water in consequence of the presence of mucilage, may be considered as yet uncertain.

The Prickly ash has a good deal of reputation in the United States as a remedy in chronic rheumatism. In that disease its operation seems analogous to that of Mezereon and Guaiacum, which it nearly resembles in its sensible properties. It is not only a popular remedy in the country, but many physicians place great reliance on its powers in rheumatic complaints, so that apothecaries generally give it a place in their shops. It is most frequently given in decoction, an ounce being boiled in about a quart of water. Dr. George Hayward, of Boston, informs me, that he formerly took this decoction in his own case of chronic rheumatism with evident relief. It was prepared as above stated, and about a pint taken in the course of a day, diluted with water sufficient to render it palatable by lessening the pungency. It was warm and grateful to the stomach, produced no nausea nor effect upon the bowels, and excited little, if any, perspiration.

I have given the powdered bark in doses of ten and twenty grains in rheumatic affections with considerable benefit. A sense of heat was produced at the stomach by taking it, but no other obvious effect. In one case it effectually removed the complaint in a few days. I have known it, however, to fail entirely in obstinate cases, sharing the opprobrium of failure with a variety of other remedies.

The Prickly ash has been employed by physicians in some cases as a topical stimulant.

It produces a powerful effect when applied to secreting surfaces and to ulcerated parts. In the West Indies much use has been made of the bark of another species, the Xanthoxylum *Clava Herculis*, in malignant ulcers, both internally administered and externally applied. Communications relating to its efficacy may be found in the eighth volume of the Medical and Physical Journal, and the fifth volume of the Transactions of the Medical Society of London.

By an ambiguity which frequently grows out of the use of common or English names of plants, the *Aralia spinosa*, a very different shrub, has been confounded with the Xanthoxylum. The Aralia, called *Angelica tree*, and sometimes *Prickly ash*, is exclusively a native of the warmer parts of the United States, being not found, to my knowledge, in the Atlantic states north of Virginia. Its flavour and pungency, as well as its general appearance, are different from those of the true Prickly ash. It is nevertheless a valuable stimulant and diaphoretic, and in Mr. Elliott's Southern Botany, we are told that it is an efficacious emetic. For the latter purpose it is given in large doses, in infusion.

The name Xanthoxylum, signifying *yellow wood*, was originally given by Mr. Colden. The spelling has since been unaccountably changed to *Zanthoxylon* in a majority of the books which contain the name. The etymology, however, can leave no doubt of the true orthography.

BOTANICAL REFERENCES.

Xanthoxylum fraxineum, SMITH, *Rees' Cycl. No.* 12.—Z. fraxineum, PURSH, i. 209.—Z. clava Herculis β. LINNÆUS, *Sp. pl.*—Z. ramiflorum, MICHAUX, *Flora*, ii. 235.—Fagara fraxini folio, DUHAMEL, *Arb.* v. *t.* 97.

MEDICAL REFERENCES.

B. S. BARTON, *Collections*, i. 25, 52; ii. 38.—THACHER, *Dispensatory*, sub Aralia spinosa.

PLATE LIX.

Fig. 1. *Xanthoxylum fraxineum in fruit.*
Fig. 2. *A barren branch in flower.*
Fig. 3. *Fertile branch in flower.*
Fig. 4. *Barren flower magnified.*
Fig. 5. *Stamen, do.*
Fig. 6. *Abortive germ of the barren flower, do.*
Fig. 7. *Fertile flower, do.*
Fig. 8. *Pistils of ditto, do.*
Fig. 9. *Perfect flower, do.*
Fig. 10. *Capsule, do. beginning to open.*
Fig. 11. *Seed, do.*

PL. LX.

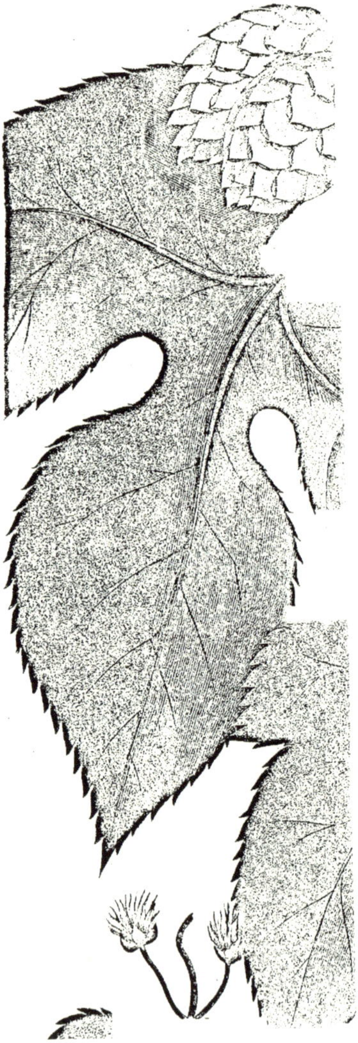

HUMULUS LUPULUS.

Common Hop.

PLATE LX.

THE Hop vine is not only a native of most countries in Europe, but is decidedly indigenous in America. It often occurs wild in the Atlantic states, and was found, by Mr. Nuttall, growing spontaneously on the banks of the Missouri. Sir J. E. Smith has quoted an old distich, which seems to be illustrative of the period of its introduction into practical use in England, about Henry the VIII's time; although he has no doubt of its being really native in that country.* The Hop being a medicinal article of some consequence, and one generally retained by the Pharmacopœias; there is a propriety in introducing it in a Medical Botany of the United States.

* "Turkeys, Carp, Hops, Pickerel and Beer
Came into England all in one year."

The genus *Humulus*, which has only a single species, is found in the Linnæan class *Diœcia*, and order *Pentandria*. It belongs to the natural orders *Scabridæ*, Linn. and *Urticæ*, Juss. Its barren flowers have a *calyx of five leaves and no corolla.* The fertile flowers have for their *calyx the scales of an ament, each two flowered; corolla of one petal, lateral; styles two; seeds solitary, invested with the corolla.*

The Hop vine is an ornamental plant, much more frequently seen cultivated than wild, and climbing to a great height. The root is perennial. Stems annual, twining from right to left, angular, rough, with minute reflexed prickles. Leaves opposite, on long winding petioles, the smaller ones heart-shaped, the larger ones three or five lobed, serrated, veiny and extremely rough. Flowering branches axillary, angular and rough. Stipules two or four, between the petioles, ovate, reflexed. Flowers numerous, and of a greenish colour. Those of the barren plants are very numerous and panicled. Their calyx has five oblong, obtuse, spreading, concave leaves. Corolla wanting. Stamens short, the anthers oblong, and bursting by two terminal pores. The fertile flowers, growing on a separate plant, are in the form of an ament, having each pair of flowers supported by a calyx-scale, which is ovate, acute, tubular at base. Corolla of one scale, obtuse, smaller than the calyx and placed one on each side of it, infolding the germ by their edge. Germ roundish, compressed; styles two, short; stigmas long, subulate, downy. The scales of the calyx and corolla swell into a kind of persistent cone or strobile, each flower producing a roundish seed.

The full grown strobiles constitute the part which is preserved for use and sold in its dried state under the name of Hops. These have an aromatic, heavy odour, and a strong, bitter, but not unpleasant taste. Besides the bitterness, they have the characteristic taste which is found in the leaves and other portions of the plant. On the outside of the scales of the calyx and corolla and near their base, is secreted a semi-resinous substance in the form of minute, yellow, transparent globules. This secretion appears to be the seat of the whole bitterness for which the hops are generally prized and consumed. Dr. Smith, in the English Botany, has observed, that the fragrance and essential properties of the hop reside in this resinous substance; and more recently an interesting series of experiments has been published by Dr. A. W. Ives, of New York, to show that this portion may effectually supercede all the rest, in common practical use.

This substance, when separated from the hops by rubbing and sifting, exists in the form of a fine yellow powder. It is adhesive when rubbed between the fingers, and becomes agglutinated by moderate heat. It is very inflammable, and burns entirely out with a white flame, leaving a light cinder.

Dr. Ives has made a variety of experiments with this powder, from which he concludes that it consists of tannin, extractive matter, a bitter principle, wax, resin, and a woody fibrous substance, besides the aromatic principle, which he was unable to separate in the form of volatile oil. It may be observed, that the powder, as employed by him, being obtained from the hops by agitation and sifting, must necessarily contain a certain portion of chaff or minute fragments of the scales; and that these are apparently the seat of the tannin, the woody insoluble substance, and possibly of some other ingredients. If the pure secretion be carefully separated from the scales by brushing, and dissolved in alcohol, it does not undergo any change of colour from the sulphate of iron; although the scales themselves, as well as the leaves of the plant, strike a black colour when treated with that salt.

Hops have long been made an ingredient in malt liquors on account of the agreeable flavour they communicate, and also from a preservative quality which they are supposed to exert in preventing acescency in those liquids. Dr. Ives has shown that a prodigious saving of expense might be made by brewers, if this powder were separated at an early period, and used instead of the hops themselves. He was able, without much trouble, to separate fourteen ounces of the powder from six pounds of hops, and concludes, that if the hops were treated, during the process of gathering and drying, with a view to the preservation of the powder, they would yield at least one pound in six. He has pointed out a vast saving, which would take place in the expense of transportation and storage, if an article containing all the strength of the hop, and occupying but small compass, were substituted for one which is of more than twenty times its bulk. An enormous loss would farther be prevented, which now takes place from the absorption produced by the hops, it being calculated that one barrel of wort is absorbed by every sixty pounds of hops used in brewing. He enumerates still farther advantages which would result from the easier preservation of the article, its superior flavour, and the diminished chance of adulteration, arising from reduction of price.*

The researches of Dr. Ives are entitled to great commendation, as they seem to promise a highly economical improvement in an important branch of domestic manufacture. In Great Britain, where malt liquors are more extensively consumed than, perhaps, in any other country, the saving must be an object of more consequence, than with us. It remains to be ascertained whether any effectual and economical method of separating the powder from the strobiles can be brought into practical use.

In medical practice the hop has been found a decided and useful tonic. A fermented decoction, known by the name of hop beer, and usually formed from this article with the simple addition of treacle, is much used in the New England states. When made sufficiently bitter with the hops, and taken as a common drink at meals, it promotes digestion more than any of

* The term *Lupulin*, by which Dr. Ives designates the powder of the hop, is convenient and not objectionable for practical use. As a chemical term, however, it does not agree with those of similar termination employed in the science; which express proximate principles of vegetables &c. and not heterogeneous bodies.

the table liquors in common use. It is serviceable in dyspeptic complaints, and is particularly adapted to obviate the lassitude and debility felt by persons of relaxed habit in the spring, or on the approach of warm weather. A simple infusion has been employed for this purpose, but the fermented liquor derives a quality from the presence of carbonic acid, which renders it more agreeable, both to the palate and stomach.

The bitter principle of the Hop, in which its tonic property appears to reside, is abundantly soluble in water. Alcohol not only extracts this portion, but dissolves also the resinous constituents of the medicine. The tincture of hops is found to be bitter and aromatic, and to exert not only a strengthening effect on the viscera, but to influence considerably the nervous system in the character of an anodyne and soporific medicine.

I have employed the tincture of hops very often in practice, and have, on the whole, had quite as much reason to be satisfied with its tonic operation, as with that of any of the bitter tinctures in common use. Its narcotic power is slight when compared with that of opium, yet it nevertheless has, in certain cases, a decided property of procuring sleep. I have particularly found it effectual in the case of persons advanced in life, who had been accustomed to the moderate; but increasing use of spirituous liquors; and who at length have considered it impossible to procure a quiet night's sleep without a preparatory draught of this kind taken warm at bed time. In such cases I have found a teaspoonful of the tincture of hops to go as far in its composing effect, as two or three ounces of ardent spirit.

Mr. Freake, who published in the Medical and Physical Journal some account of the properties of this medicine, states that he had found it decidedly advantageous in erysipelas, in gout and in some other diseases. He considers its beneficial effects to arise from its alterative and tonic power on the system. He thinks it sedative, aperient and diuretic; and a good antiseptic and corroborant in bowel complaints. In his practice he had found pain to be eased and rest procured with this medicine, when opium did not succeed.

Dr. Maton found that besides allaying pain and procuring sleep, the preparations of hops were capable of reducing the frequency of the pulse, and increasing its firmness in a direct manner. One drachm of the tincture and four grains of the extract given once in six hours reduced the pulsations from ninety six to sixty in the course of twenty four hours. He found the extract very efficacious in allaying the pain of of articular rheumatism.

Some experiments of Dr. Bigsby and others have not been found to confirm the previous character of this article in all the foregoing respects, and its sedative powers have been called in question. As in most new medicines, its virtues have doubtless been exaggerated by its earliest advocates; yet it is not on this account to be discarded from use. Although the narcotic powers of the hop are not of the most energetic kind, they nevertheless do exist, and the very circumstance of their mild and temperate influence renders them, in many cases, safer than those of more active drugs.

In regard to the lithontriptic power which has been imputed to hops both alone, and through the medium of malt liquors, it is not probable that they have any operation of this sort, beyond that of a palliative.

The external application of hops has long had the popular reputation of being anodyne and composing. A pillow of hops is thought instrumental in procuring sleep, but with what justice I am unable to say. Poultices and fomentations made of them are in repute as sedative applications for painful swellings. When steeped in hot brandy and held in the mouth, they sometimes relieve the pain of a carious tooth. For all these purposes, no doubt, they often fail, yet there is little temerity in asserting that they are equally to be depended on in such cases, with the rest of the articles of the Materia Medica.

The most common form for internal use, where a sedative effect is desired, is that of the saturated tincture. The powder separated from the hops may be given in substance with a certainty of securing all their medicinal effects. This powder must be given in small doses, to be retained on the stomach and bowels. Dr. Bryorly found that twenty or twenty five grains left a sense of acrimony in the throat, and were followed by a good deal of nausea, and in some instances by purging.

The vine of the hop has been appropriated to some economical uses. In spring, when the young shoots first emerge from the ground, they are boiled and eaten as asparagus, and are accounted very salubrious. The fibres of the vine are strong and flexible and have been manufactured into a coarse cloth in Sweden and England, particularly for the sacks in which the hops are carried to market.

BOTANICAL REFERENCES.

Humulus lupulus, LINN. *Sp. pl.*—SMITH, *Engl. Bot. t.* 427.—MILLER, *Illustrations, t.* 88.—MICHAUX, ii.—PURSH, i. 199.—NUTTALL, ii. 237.—Lupulus mas et fæmina, RAY, *Syn.* 137.

MEDICAL REFERENCES.

FREAKE, *Med. and Phys. Journal,* xiii. 432.—THOMPSON, *London Dispensatory,* 200.—BIGSBY, *London Medical Repository,* v. 97.—BRYORLY, *Inaug. Diss. Philad.* 1803.—IVES, *in Silliman's Journal,* ii. 302.

PLATE LX.

Fig. 1. *Humulus lupulus in fruit.*

Fig. 2. *Fertile flowers.*

Fig. 3. *Calyx and pistil of do. (the corolla omitted by mistake of the engraver.)*

Fig. 4. *Barren flowers.*

Fig. 5. *Stamen magnified.*

NOTES.

Note A.

Ranunculi, quod aliqui apium agreste nominant, plura quidem sunt genera: at vis tamen omnibus una, acris scilicet ac vehementer exulcerans. Ac unum quidem coriandri foliis constat, sed latioribus, subalbidis et pinguibus: flore luteo, interdùm purpureo. Caulis minimé crassus est, sed cubitum altus. Radice nititur exigua, candida et amara, adnatis ceu capillamentis, hellebori modo, fibrata: juxta fluenta nascitur. Alterum est lanuginosius, longioréque caule, pluribus foliorum incisuris, plurimum in Saridinia proveniens, acerrimum, quod etiam sylvestre apium appellant. Est et tertium valde parvum et odore gravi; flore aureo. Quartum simile huic, flore lacteo. Folia et caules tenelli vim habent illitu exulcerandi, et usque adeò urendi, ut etiam crustas cum dolore inducant. Quare scabros ungues auferunt, psoras removent, stigmata delent: itemque formicationes ac pensiles verrucas et alopecias ad breve tempus imposita tollunt. Quin et repente eorum decocto perniones foventur. Radix verò sicca tritáque, sternutamenta ciet, naribus admota: dentium quoque dolores appensa levat, ipsos tamen rumpit.

Dioscorides. Interp. Sarraceni, Lib. ii. Cap. 206.

Note B.

Sponte jam patet, internum Euphorbii usum periculo plenum esse. Sed confirmant id infortunia, specialibus casibus subnata. Obiit quidam, cui empiricus illud imprudenter exhibuerat, dysenteria eodem die. Virgo venusta seni decrepito, se invita, desponsata ad mortem sibi conciliandam pulverem Euphorbii ingessit, unde dolores ventris atrocissimi, hypercatharses cum vomitionibus frequentissimis, singultu, ardore ventriculi et faucium sitique inextinguibili, tandem sudores frigidi et animi deliquia: ex quibus angustiis tamen arte emersit. Nihilomimus tamen quidam illud præcipere ausi sunt, et instar drastici, quod pituitam, sed potentius aquam, subduceret in iis, quibus venter nimis contra alia mitiora torpet, vel ut loqui amant, friget, in hydropicis præcipue admiserunt. Ita Ætius, Actuarius et Arabes non nulli. Galenus et Dioscorides tacent de vi ejus purgante. Omnes tamen, qui ore captum concedunt, cautionem summam injungunt, et connubium, cum iis, quæ acrimonium ejus frangere valent, vel prægressam mitigationem desiderant. Hanc ipsam tentarunt oleo amygdalino, succo Citri, phlegmate Vitrioli, Mastiche, Croco, Tragacantha, Melle aliisque bene multis secundum varium de ejus natura conceptum. Sed ejusmodi correctiones vel non sufficiunt, vel, si sufficiunt, ipsam vim medicaminis destruunt. Minuere dosin vel rite illam diluere, aliis exemplis artis est. Ast nondum comprobata vera dosis est. Ad grana decem permittit. Sennertus, alias non ineptus subdolæ Euphorbii vis judex; a grano uno ad octo cum semisse concedit Heurnius; a granis duobus sex vel octo Geoffroy. Omnibus hisce audacior et Fallopius, qui præceptoris sui Machesii auctoritate et propria experientia ductus, non dubitavit Euphorbii vetusti drachmam unam, rarius scrupulos quatuor, dare. Mixtum Cassia mitius deprehendit quam solutum, qua forma sitim intolerabilem et evacuationem largiorem creavit. Sed praestat usum internum eiusdem omnino negligere.

Murray Apperatus Medicaminum sub Euphorbia officinarum.

Note C.

Nymphæa in paludibus stagnantibúsque aquis nascitur: folia veró habet Ægyptiæ fabæ similia, at minora oblongioráque, plura ab una eadémque radice prodeuntia: quorum alia super aquam quodammodo extant, alia in ea ipsa demerguntur: florem album, lilio similem, in quo medium croceúm est. At cúm defloruerit, calyculus rotundus, figura malo aut papaveris capiti similis, idemque niger, extuberat: in quo semen nigrum, latum, densum, atque gustanti lentum glutinosumve recluditur. Caulis est lævis, minimé crassus, niger, Ægyptiæ fabæ cauli similis: radix nigra, scabra, clavæ similis, quæ autumno secatur. Ea sicca, cum vino pota, cœliacis ac dysentericis auxiliatur, lienemque consumit. Stomachi quoque ac vesicæ doloribus sedandis ipsa radix imponitur, et alphos ex aqua emendat: alopeciis etiamnum cum pice imposita medetur. Eadèm contra veneris insomnia bibitur, siquidem illa in totúm adimit: quin et aliquot diebus continenter epota, genitale ita infirmat, ut arrigi minimé possit. Idem porro seminis quoque poti effectus est. Cæterùm à nymphis nymphææ nomen sibi vendicasse creditur, quoniam loca amet aquosa. Plurima autem inuenitur in Helide, Anygro amne, et in Bœotiæ Aliarto.

Dioscorides interp. Sarraceni, iii 148.

Spotted something that isn't quite right?

Let us know and we'll fix it for future reprints.
And as a thank you for your diligence, we'll
send you a free Sothis book of your choice.

publisher@sothis.press

This page intentionally left blank

Made in United States
Orlando, FL
04 November 2026

90206686R00388